THE CURIOUS HISTORY
OF THE HEART

A Cultural and Scientific Journey

VINCENT M. FIGUEREDO

Columbia University Press

New York

Columbia University Press
Publishers Since 1893
New York   Chichester, West Sussex
cup.columbia.edu

Library of Congress Cataloging-in-Publication Data
Names: Figueredo, Vincent M., author.
Title: The curious history of the heart : a cultural and scientific
journey / Vincent M. Figueredo.
Description: New York : Columbia University Press, [2023] |
Includes bibliographical references and index.
Identifiers: LCCN 2022023161 | ISBN 9780231208185 (hardback) |
ISBN 9780231557306 (ebook)
Subjects: LCSH: Heart—Popular works. | Cardiology—History.
Classification: LCC QP111.4 .F54 2023 |
DDC 612.1/7—dc23/eng/20220906
LC record available at https://lccn.loc.gov/2022023161

Cover design and illustration: Henry Sene Yee

To the women of Five Fig Farm:
Ann, Sarah, Isabel, and Madeline

You fill my heart with love

CONTENTS

CONTENTS

CONTENTS

THE CURIOUS HISTORY
OF THE HEART

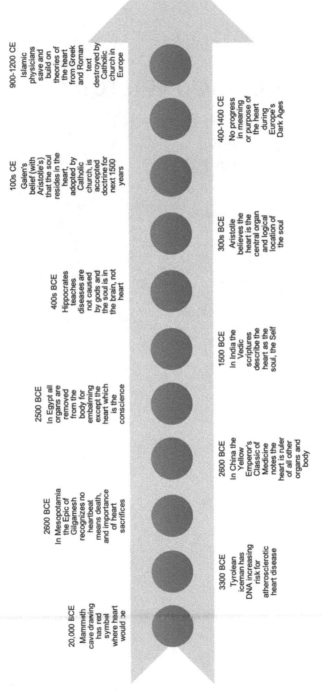

FIGURES 0.1–0.3 Timeline of the heart through human history.

Source: Created by author.

20,000 BCE
Mammoth cave drawing has red symbol where heart would be

2600 BCE
In Mesopotamia the Epic of Gilgamesh recognizes no heartbeat means death, and importance of heart sacrifices

2500 BCE
In Egypt all organs are removed from the body for embalming except the heart which is the conscience

400s BCE
Hippocrates teaches diseases are not caused by gods and the soul is in the brain, not heart

100s CE
Galen's belief (with Aristotle's) that the soul resides in the heart, is adopted by Catholic church, is accepted doctrine for next 1500 years

900-1200 CE
Islamic physicians save and build on theories of the heart from Greek and Roman text destroyed by Catholic church in Europe

3300 BCE
Tyrolean iceman has DNA increasing risk for atherosclerotic heart disease

2600 BCE
In China the Yellow Emperor's Classic of Medicine notes the heart is ruler of all other organs and body

1500 BCE
In India the Vedic scriptures describe the heart as the soul, the Self

300s BCE
Aristotle believes the heart is the central organ and logical location of the soul

400-1400 CE
No progress in meaning or purpose of the heart during Europe's Dark Ages

1100's
Royalty have their hearts, their spiritual and moral center, buried separate from bodies at their favorite place of worship

1200's
Vikings believe the smaller and colder a heart, the braver the warrior

1400s
da Vinci draws first anatomically correct heart and makes new heart discoveries, unfortunately not rediscovered for 150 years

1600s
Harvey first describes the heart as a pump and the circulation of blood

1800s
Laennec discovers the stethoscope

1929
Forssmann performs first human heart catheterization on himself

1200s
Reports of saints having inscriptions on the inner walls of their hearts about their love for God and Jesus

1300-1500s
Aztecs cut out beating hearts by the thousands to help the god Huizilopochtii fight off darkness and the end of the world

1500s
Vesalius makes an art of body snatching, drawing the first anatomically correct heart published

1700s
Heberden calls crushing chest sensation with exercise 'angina pectoris'

1896
Rehn performs first heart surgery on 22-year-old stabbed in the heart, sewing hole with catgut

FIGURES 0.1–0.3 (*continued*)

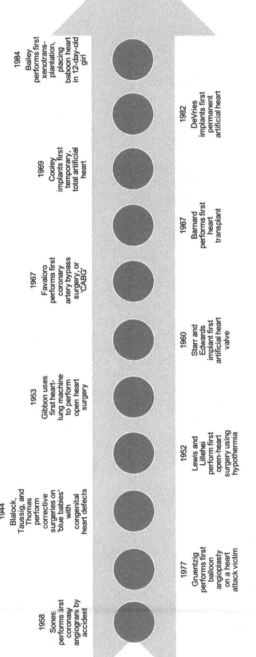

1958
Sones performs first coronary angiogram by accident

1977
Gruentzig performs first balloon angioplasty on a heart attack victim

1944
Blalock, Taussig, and Thomas perform corrective surgeries on 'blue babies' with congenital heart defects

1952
Lewis and Lillehei perform first open-heart surgery using hypothermia

1953
Gibbon uses first heart-lung machine to perform open heart surgery

1960
Starr and Edwards implant first artificial heart valve

1967
Favaloro performs first coronary artery bypass surgery, or 'CABG'

1967
Barnard performs first heart transplant

1969
Cooley implants first temporary, total artificial heart

1982
DeVries implants first permanent artificial heart

1984
Bailey performs first xenotransplantation, placing baboon heart in 12-day-old girl

FIGURES 0.1–0.3 (*continued*)

INTRODUCTION

KING CHARLES I of England reached forward and placed his three fingers and thumb into the gaping hole in the young nobleman's left chest. The king gently touched this man's beating heart.

"Does it hurt?" he asked.

"Not at all," said the young man.

It was 1641 CE, and Charles had heard of this miracle from his personal physician, William Harvey, who was the first to scientifically demonstrate the heart's role in the circulation of blood throughout the body. With great interest the king had asked if he could meet this young man who was the nineteen-year-old son of the Viscount of Montgomery in Ireland.

At age ten, the boy fell from a stumbling horse onto a jutting rock, which pierced and broke multiple ribs on his left side. The wound abscessed and healed, leaving a hole in the boy's left chest. Nine years later, alive and well, the now famous nobleman had just returned to London after traveling the European continent to sellout crowds who wanted to witness a beating heart in a living person. After Harvey examined the young man with the king, he wrote: "I have handled the heart and ventricle, in their own pulsations in a young and sprightly nobleman without offence to him; wherefore, I conclude that the heart is deprived of the sense of feeling."[1]

It is a great irony that the heart—which throughout history has been placed at the center of human feeling—is in fact numb to physical touch. From the time humans first recorded their thoughts, most civilizations believed that the heart, not the brain, was the most important organ in the body. Surely ancient humans knew that the beating in their chest signified life—beating harder and faster with fear or desire, and upon death beating no more. For thousands of years, the Egyptians, Greeks, Chinese, and Teotihuacans of Mesoamerica elevated the heart to the position held today by the brain: as the seat of the soul, emotions, thoughts, and intelligence. Throughout history, many societies thought it was through their heart that a person connected with God; and God measured a person's chances of eternal heavenly bliss based on the virtues and sins of one's life recorded in the walls of the heart.

Harvey's observation in 1641 CE that the heart acted as a circulation pump had ramifications for centuries to come. Scientists and physicians changed their beliefs about the heart, and the brain slowly took over as the director and sole repository of emotions and consciousness. Today most of us believe that our brain controls our body, including the function of the heart. The heart, we've been taught, is just a pump pushing blood around the body through a circulatory system.

Because we have come to accept the heart as nothing more than a pump, we have decided that it is morally sound to transplant the heart of one person into another. But every once in a while a case arises like that of Claire Sylvia. A former professional dancer, she underwent a heart-lung transplant, receiving a heart from eighteen-year-old Tim Lamirande, who died in a motorcycle accident. After her heart transplantation, friends remarked that Claire started walking like a man; she began craving beer and chicken nuggets, which she detested prior to her transplant. Tim's family said these were his behaviors. They were not surprised that she was acting this way because she now had Tim's heart inside her. This story was the subject of the 2013 movie *Heart of a Stranger* starring Jane Seymour, but multiple accounts of inheriting the personality traits of a donor after heart transplantation have been recorded. These stories make us wonder whether the heart is just a mechanical pump or whether emotional parts of us are contained in it and travel with it.

As a cardiologist, I regularly encounter cases in which the emotional and physiological heart show a profound connection. I have witnessed heart attacks after the sudden loss of a loved one when the patient had no prior heart disease. Other patients experienced a heart attack or sudden death after their team lost a Super Bowl or World Cup penalty shootout. I have frequently witnessed lifelong couples dying within months of each other. Despite these many cases and the millennia-old association of the heart with emotions, modern medicine seems to have dismissed this intimate connection. In this book, I recount the history explaining how this happened and reveal how modern science is now suggesting that what we've lost to history should be reconsidered.

Medical science recently has found that the heart may hold feelings and, in fact, is part of a two-way "heart-brain connection." Studies suggest that the heart directs the brain as much as the brain directs the heart.[2] New research in this area may be the beginning of a scientific shift, aligning historic and modern cultural views of the heart. The heart may no longer be viewed merely as a pump; rather, the heart may again be recognized as part of the emotional vitality that ensures our mental, spiritual, and physical health.

The heart is the first organ to react to signals from the brain—think fight or flight. When a mountain lion appears on your walk in the woods, the brain activates the sympathetic nervous system, triggering an acute response that prepares the body to stand and fight or to flee. The brain tells the heart to immediately beat faster and stronger, pushing oxygenated blood to the body's muscles to prepare them to move. The brain is also first to receive signals from the heart. If that weren't the case, we might pass out when we quickly stand. The heart and its great vessels alert the brain that blood volume and pressure are dropping, and the brain responds by activating blood vessels to constrict to prevent blood from pooling in the legs.

The emotions we register in our brain reverberate in our heart. The resulting physical sensations of seeing a new love—blushing, warmth, racing pulse—are manifestations of the heart's response. It is this interdependence, this heart-brain connection, that is so vital to our health. It's what has led humans for thousands of years to place their emotions, reasoning, and very soul in this hot, pumping organ that signifies we're alive. The ancient Chinese and Indians stressed that a happy heart meant a happy body and a long, healthy life. The brain was seen as a mass of cold gray pudding, nothing more than a phlegm-producing organ, and the ancient Egyptians pulled it out through the nose with a hook during the embalming process.

Today the brain has taken over as the home of our consciousness, but the heart continues to play a central role in our cultural iconography. You only have to see the emojis on your phone in a text from your love, or the heart emblem on a car bumper sticker, to realize the important role the heart now plays in our lives, at least symbolically. The heart remains a symbol of romance and love, and more recently the heart ideograph has become familiar to us as a sign of health and life.

We still say, emotionally, "I love you with all my heart," "you have touched my heart," and "she broke my heart." We declare, "he's heartless." We beg others to "please have a heart." "She speaks from the heart" connotes sincerity and honesty. A "change of heart" implies reconciliation or repentance. Referring to our intelligence, "we memorize it by heart." What part of the body do we point to when we say "me"? Yet modern medicine has rejected the heart as the repository of our soul, intelligence, and feelings. We have mostly forgotten the place of the heart in our past, even though this important role remains pervasive in our inherited cultural icons, poetry, and art.

Despite advances in medicine, one out of three of us will die from heart disease worldwide. More of us will die of cardiovascular disease than from all cancers combined. Heart disease kills ten times more women than breast cancer. In the United States, someone dies of a heart attack every forty seconds. Why are three of the biggest drivers of our current health crisis—heart disease, depression, and stress—not considered and treated in a more connected way?

More than any other area of medicine, cardiology was at the forefront of innovation in the twentieth century and is even more so in the twenty-first century. The twentieth century saw development of coronary artery bypass surgery, catheter-based coronary balloon angioplasty and stents, pacemakers and defibrillators, heart assist devices, and heart transplants. Preventive health measures directed at cardiac risk factors such as smoking, high blood pressure, and cholesterol—half of Americans currently have at least one of these cardiac risk factors—have helped to decrease deaths due to heart disease. As a result, the incidence of cardiovascular disease has decreased significantly since the 1960s—yet it remains the number one killer of us all.[3]

I believe that part of the answer to improving our collective heart health lies in a better understanding of the cultural and scientific history of the heart and how it became separated from and dominated by the brain. Today the heart is a "replaceable" organ. If a donor heart is not immediately available, a mechanical pump can be implanted into the chest to replace the heart's function while the patient with a failing heart waits. Scientists are now looking at growing a whole new three-dimensional heart from one's own cells to replace the failing heart. Ongoing studies are investigating placing the heart of another animal, such as a pig, into a human because of the shortage of available donor human hearts.[4] And soon gene-informed personalized medicine will enable each of us to be assessed and treated for heart disease based on our unique genetic risks.[5]

I have spent most of my life studying and caring for hearts. This compelled me to take a panoptic look at the meaning of the heart over the whole of human history. I explored how the battle between the heart and the brain has led to our current cultural and scientific understanding of the heart-brain connection. In this book, I trace the evolution of our understanding of the heart from the dawn of human civilization twenty thousand years ago to today (see figures 0.1–0.3). I examine how our beliefs about the purpose of the heart have evolved and how that affects our understanding of what life forces it contains. We have always considered our heart to be at the center of our body—the center of the body is actually below our belly button near our sacrum—but what do we believe the heart is central to?

I looked back to see what our ancestors thought of this wondrous organ: the heart has been revered, regaled, misunderstood, and revealed through the ages. Throughout history the heart has played an important role for the poet, the philosopher, and the physician. The heart has meant different things in different cultures, from prehistoric humans, through ancient societies, the Dark Ages, the Renaissance, and into modern times. I chronologically examine how the "king" of the organs became dismissed as a mere mechanistic blood pump subservient to the brain even as it remained central to our daily lives as a symbol of love and health. As a physician fascinated with this extraordinary organ, I include a section on how the heart works and heart diseases. In addition, I discuss advancements in heart therapies in modern times and what the future may hold. What we are now learning shows that our ancient ancestors weren't so wrong about the heart after all.

I have "poured my heart and soul" into this book. I hope you find this curious history of the heart to be as fascinating as I have.

PART 1

THE ANCIENT HEART

Chapter One

THE HEART MEANS LIFE

IN 1908, archeologists discovered a painting of a mammoth with what looks like a red heart drawn on its chest, on the wall of El Pindal cave in Asturias, Spain (figure 1.1). Drawn by the Magdelenian people during the Upper Paleolithic, the painting dates from 14,000 to 20,000 years ago. The ancient artist may have known that the best way to kill an animal was to hit it through this red, beating organ. The image may have been painted as a target.

By 12,000 years ago, when humans began to settle into villages, towns, and city-states, they probably already believed that the heart was the most important organ in their body—the reason they were alive.

■ ■ ■

FIGURE 1.1 Mammoth with what appears to be a heart drawn as a target.
El Pindal cave, Asturias, Spain.
Album / Art Resource, New York.

I touch his heart but it does not beat at all.

Epic of Gilgamesh, Tablet 8, 2600 BCE

Gilgamesh, the hero king of the ancient Mesopotamian *Epic of Gilgamesh*—the oldest written story known to exist—uttered this lament on the death of his friend Enkidu.[1] Gilgamesh, king of Uruk (a city-state in ancient Mesopotamia), and Enkidu were initially enemies, but they grew to respect each other and eventually became best friends. After meeting Enkidu, Gilgamesh became a better king to his people because he understood them better. Enkidu was killed by the gods in revenge for helping Gilgamesh slay the divine bull sent by the goddess Ishtar to destroy Gilgamesh (for rejecting her salacious advances).

Gilgamesh tries to revive his friend only to find his heart beats no more. Written in what is now modern Iraq in Sumerian cuneiform around 2600 BCE, this passage may be the earliest reference to pulse-taking.[2] More than 4,600 years ago humans understood that our hearts beat and the pulsations could be felt throughout the body. After killing the divine bull, Gilgamesh and Enkidu cut out its heart as an oblation to the sun god Shamash—the first heart sacrifice ever recorded. As in many ancient societies, the heart occupied an important place in Sumerian culture. It was the primary organ of life in the body and the choice sacrifice to appease the gods.

In 1849, Sumerian medical tablets from as early as 2400 BCE were discovered in the ancient Assyrian cities of Assur and Ninevah. Most of these medical texts came from the Royal Library of Ashurbanipal (600s BCE). Ashurbanipal was considered the last great king of Assyria. The Epic of Gilgamesh was found in his library.

Mesopotamians knew little about anatomy and physiology because they were restricted by religious taboo from performing human dissection. Their approach to diseases and death were spiritual rather than physiological or anatomical. The heart was believed to be the location of intellect, the liver of affectivity, the stomach of cunning, and the uterus of compassion. There is no mention of the brain in their medical literature. Neurological and psychiatric disorders, such as epilepsy, strokes, depression, and anxiety, were believed to be caused by angry gods and demons attacking the unfortunate person and were dealt with by religious healers from the temples. These healers' duties mostly consisted of exorcisms, addressing the spirit whose malevolent activities were causing the patient's symptoms. The cuneiform texts suggest healers did observe clinical symptoms and gave herbal medicines, such as to relieve pain. Sumerian healers also felt patients' pulses to assess their health. Enkidu had no pulse, and therefore, no life.

Other civilizations advancing elsewhere simultaneously, such as the Egyptians and Chinese, developed their own ideas about the purpose and importance of the heart. All of these ancient civilizations agreed on one thing: the beating heart meant life.

■ ■ ■

O my heart which I had from my mother, O my heart which I had upon earth, do not rise up against me as a witness in the presence of the Lord of Things; do not speak against me concerning what I have done, do not bring up anything against me in the presence of the Great God, Lord of the West.[3]

An Egyptian in 2500 BCE believed that Anubis, the god of the dead (the jackal headed god, as jackals hung around cemeteries), would take the person who died into Duat, the underworld. The person was presented to Osiris, god of the underworld and afterlife, and to a tribunal of forty-three deities in the Hall of

Maat, goddess of justice. There, the person's heart was weighed on the scale of justice against the Feather of Maat—an ostrich feather that represented truth (figure 1.2). If the heart was lighter or equal to the weight of the feather, the person had led a virtuous life, and Osiris escorted the person to the Field of Reeds, a heavenly paradise. If the heart weighed more than the feather, the goddess Ammit (part crocodile, lion, and hippopotamus) ate their heart, and the person's soul vanished from existence.

FIGURE 1.2 The Weighing of the Heart from the Book of the Dead of Ani. On the left, Ani and his wife Tutu enter the assemblage of gods. In the center, Anubis weighs Ani's heart against the feather of Maat, observed by the goddesses Renenutet and Meshkenet, the god Shay, and Ani's own ba. On the right, the monster Ammit, who will devour Ani's soul if he is unworthy, awaits the verdict, and the god Thoth prepares to record it. Across the top are gods acting as judges: Hu and Sia, Hathor, Horus, Isis and Nephthys, Nut, Geb, Tefnut, Shu, Atum, and Ra-Horakhty.

Source: British Museum / Wikimedia Commons / Public Domain.

Ancient Egyptians believed the heart was witness to everything people did in life, good or bad. Because many did not lead the most virtuous life, they were worried that their heart might testify against them. The heart could become weighted down with sin. To prevent the heart from testifying against them, at death a heart scarab would be wrapped within the bandages on their chest when the body was prepared for mummification. The inscription on the scarab comes from Chapter 30 of the Book of the Dead (the quote at the beginning of this section).

The heart was the source of life and being, and ancient Egyptians treated it with reverence during embalming rituals. It was important that the heart remain within the body as it traveled to the underworld to be judged by Osiris. The heart was the only organ to be replaced in the body after embalming. The other chest and abdominal organs were placed in jars near the mummy. But the brain was considered worthless, doing nothing more than passing mucous to the nose. The ancient Egyptian word for *brain* roughly translates to "skull offal." So while the heart was carefully preserved and placed back in the body, the brain was scraped out of the skull with an iron hook through the nose and discarded.

It is unclear where the practice of medicine began: some believe in ancient Egypt, others in ancient Mesopotamia. Papyri from as long ago as 1950 BCE suggest that Egyptians were practicing medicine and studying the heart more than 4,000 years ago.[4]

The first physical descriptions of the heart and how it functions are found in three Egyptian medical papyri: the Edwin Smith Papyrus (c. 1500 BCE; the world's first surgical text), the Ebers Papyrus (c. 1550 BCE), and the Brugsch Papyrus (c. 1350 BCE). These ancient Egyptian medical texts are dated from a later period than Mesopotamian medical cuneiform tablets (c. 2400 BCE). However, these Egyptian papyri are believed to be copies of much older texts from as early as 2700 BCE, possibly from the writings of Imhotep, a high priest and physician of the Old Kingdom. Imhotep, chancellor to the pharaoh Djoser, was the architect of the Djoser step pyramid and may have been the first in history to use stone columns to support a building. He was the chief physician to Djoser. Imhotep apparently wrote extensively on architecture and medicine. It is thought that he is the author of the material upon which the later medical papyri are based, especially the Smith Papyrus. He was deified as the Egyptian god of medicine and healing 2,000 years after his death—one of a few nonroyal Egyptians to become a god.

Because the Egyptians practiced embalming procedures for mummification, it makes sense that they had an intimate knowledge of anatomy. Ancient Egyptian doctors believed that the heart gave rise to vessels that could be distally palpated. In the Ebers Papyrus, it is written that "the heart speaks out of every limb":

From the heart arise the vessels which go to the whole body . . . if the physician lay the hands or his fingers to the head, to the back of the head, to the hands, to the place of the stomach, to the arms or to the feet, then he examines the heart, because all of his limbs possess its vessels, that is: the heart speaks out of the vessels of every limb . . . if the heart trembles, has little power and sinks, the disease is advancing.

For ancient Egyptians, the heart was the center of the body, with vessels attached to every part. When a person fainted, they observed that the pulse temporarily disappeared. They described a weak pulse with displacement of the heart impulse on the chest to the left of its usual position; what we now recognize as a weak, enlarged heart consistent with congestive heart failure. Excessive salivation was called "flooding of the heart," a possible description of a person in acute heart failure with excessive pink (blood tinged) frothy sputum. In the Ebers Papyrus, ancient Egyptians reported that pains in the arm and the breast on the side of the heart suggested that death was approaching. This is a classic description of a heart attack!

Ancient Egyptians ascribed intellect to the heart, commanding all other organs. The heart was necessary to keep the body working, to keep it alive, as explained in the Ebers Papyrus:

> It is the fact that the heart and tongue rule over the limbs, according to the doctrine that the heart is in each body and the tongue in each mouth of all Gods, men, and beasts. Inasmuch as the heart thinks whatsoever it will, and the tongue commands all that it wills. The seeing of the eyes, the hearing of the ears, the breathing of the nose, they bring tidings to the heart. It is the heart which brings into being each act of intelligence, and the tongue it is which repeats what is thought by the heart. And so, all works are performed, and all handiwork, the doing of the hands, the going of the feet, the movement of all the limbs, according to its command.

As the center of their being, ancient Egyptians believed the heart moved not only blood but air, tears, saliva, mucus, urine, and semen through the body by a system of channels. The heart sustained life and meant life.

Thus, ancient Mesopotamians and Egyptians had decided the heart was the most important organ in the body. Its beating meant life. It had to remain with the body to go to the afterlife. Meanwhile, ancient Chinese were also studying the body. They came to believe that the heart ruled the body as a king.

■ ■ ■

For ancient Chinese, the heart was the king of all organs.[5] All other organs sacrificed for the heart; they gave their energy to help the heart maintain its balance. The ruling heart was responsible for maintaining internal peace and harmony through the entire body. The heart was the force responsible for physical, mental, emotional, and spiritual well-being. Based on his study of the ancient medical book *Huangdi Neijing* from 2600 BCE, Guan Zhong in his Daoist classic Guanzi wrote prior to the third century BCE:

The heart is the emperor of the human body. Its subordinate officers are in charge of the nine orifices [two eyes, two ears, two nostrils, one mouth, one urethra, one anus] and their related functions. As long as the heart remains on its rightful path, the nine orifices will follow along and function properly. If the heart's desires become abundant, however, the eyes will lose their sense of color, and the ears will lose their sense of sound.[6]

Forty-seven centuries ago, *Huangdi Neijing* (The Yellow Emperor's Classic of Medicine) was written by the Chinese emperor Huangdi. This book records discussions between the emperor and his physician in which Huangdi questions him about the nature of health, disease, and treatment. Expanding on Huangdi's belief that the heart rules the five organ networks, scholars from the court of Liu An, king of Huainan, in the second century BCE, wrote in the Daoist classic *Huainanzi*:

> The heart is the ruler of the five organ networks. It commands the movements of the four extremities, it circulates the qi [life energy] and the blood, it roams the realms of the material and the immaterial, and it is in tune with the gateways of every action. Therefore, coveting to govern the flow of energy on earth without possessing a heart would be like aspiring to tune gongs and drums without ears, or like trying to read a piece of fancy literature without eyes.[7]

Huangdi Neijing was viewed as the most important text of ancient Chinese medicine, and it has remained a reference work for practitioners of traditional Chinese medicine well into the modern era. This early medical text was concurrent with Sumerian medical tablets (2400 BCE) and the Egyptian physician Imhotep (2700 BCE). All three cultures viewed the heart as the most important organ in the body; the ruler of the body and determinant of life. During the Ming Dynasty, in 1570 CE Li Yuheng wrote:

The ancient book of definitions [*Huangdi Neijing*] refers to the heart as the ruler of the human body, the seat of consciousness and intelligence. If we decide to nourish this crucial element in our daily practice, then our lives will be long, healthy, and secure. If the ruler's vision becomes distracted and unclear, however, the path will become congested, and severe harm to the material body will result. If we lead lives that are centered around distracting thoughts and activities, harmful consequences will result.[8]

Another example of the lasting effect of *Huangdi Neijing* on the Chinese view of the heart's life importance can be seen in the writings of Li Ting in 1575 CE:

The heart is the master of the body and the emperor of the organ networks. There is the structural heart made from blood and flesh: it has the shape of a closed lotus flower and is situated underneath the lung and above the liver. And there is the luminous heart of spirit which generates qi and blood and thus is the root of life.[9]

Four thousand years before William Harvey "discovered" the circulatory system, it seems that Chinese physicians understood blood circulation. Passages from the *Huangdi Neijing* include "All blood is under the control of the heart"; "The blood flows continuously in a circle and never stops"; and "the blood [qi] flows but continuously like the current of the river, or the sun and moon in their orbits. It may be compared to a circle without beginning or end."[10]

For ancient Chinese, the heart ruled the body's organs. It was the "root of life" generating spirit and blood to nourish the body. A happy heart meant a healthy life. The brain was nothing more than nourishing marrow (like bone marrow). What we now accept as functions of the brain were ascribed to the five *zang* organs: from the heart, liver, spleen, lung, and kidney arose happiness, anger, deep thinking, melancholy, and fear, respectively.

■ ■ ■

Like the ancient Chinese, ancient Indians also believed the heart was the seat of life and consciousness. Ayurvedic medicine, one of the world's oldest holistic (whole body) healing systems, described the heart as the prime mover of *prana* or life force.[11] Ayurveda is based on the belief that one's health depends on a balance between mind, body, and spirit. The ancient writings on Ayurvedic medicine come from the early Vedic era in India around 1600 BCE.

Ayurvedic medical knowledge was documented in compendia known as Samhitas, found in the four Vedas, which are the oldest holy books in Hinduism. The Charaka Samhita (500 BCE) described what was understood about the human body, diet and hygiene, and symptoms and treatments for a wide range of diseases. The Sushruta Samhita (200 BCE) described cadaveric dissection, embryology, and human anatomy. There is even a section on treating alcoholism (yes, it was a problem in 200 BCE as well).

Ayurvedic Samhitas described the heart as having ten prominent outlets or vessels, which is similar to the nine gates described by ancient Chinese medical practitioners. The vessels arising from the heart transported nutrients to the rest of the body. The heart fed the body *rasa vaha srotas* or the "juice of life."

Mana or mind was located in the heart. Mana coordinated the sensory organs, the organs of action, and the soul. In the Charaka Samhita, the mind and thought resided in the heart. In the Sushruta Samhita, it is stated that the heart in the embryo is developed first as the seat of the mind and intellect. Although it was generally accepted in early Ayurvedic teachings that the heart was home to the soul and consciousness, some challenged this traditional thinking. In the Bhela Samhita (c. 400 BCE), it was written that mana was in the head, whereas *chitta* or thinking was in the heart. The motor and sensory functions of the mind were attributed to the brain, but the psychological functions were attributed to the heart. It may be that these early Ayurvedic thinkers were describing the heart-brain connection over two thousand years ago.

Alexander the Great and his army, which included scholars and physicians, peaceably gained control of Taxila (in what is now Pakistan), situated at the junction of the Indian subcontinent and central Asia in 326 BCE. The mixing of these two cultures meant that interactions took place between the academics of ancient India and ancient Greece. There are, in fact, striking similarities between theories on the heart in these two systems of medical knowledge.

■ ■ ■

The early Greeks believed the heart was determinate of life: whether a human or a god. Dionysus, the god of wine and ecstasy—worshiped as early as 1500s BCE by Mycenean Greeks—was the child of Zeus and Persephone. Jealous Hera, Zeus's wife, had the Titans kill the child. They cut him to pieces and boiled the parts to be eaten. Athena, who was Zeus's favorite daughter (born out of his head full-grown and clothed in armor), managed to save Dionysus's heart before the Titans ate it. Zeus then ground up Dionysus's heart, placed in a potion, and gave it to a beautiful mortal princess, Semele, to drink. She burnt to a crisp when she asked Zeus to appear before her in his true form, but not before Zeus rescued Dionysus from her womb and sewed him into his thigh until he was born.

Despite similarities to ancient Greek theories of the heart, Ayurvedic medicine may have gone a step further, describing that rasa reenters the heart once again, after being carried to all parts of the body (that would be the concept of circulation, predating Harvey's discovery by two thousand years). In the Bhela Samhita (c. 400 BCE), "The blood [rasa] is first ejected out of the heart, it is then distributed to all parts of the body, and thereafter, is returned to the heart."

Although there may have been some interchange of knowledge between ancient India and Greece, the Greeks—and subsequently the Romans—held stubbornly to their own theories on the workings of the heart and the body. When Europe fell into the Dark Ages for a thousand years, there was a prohibition on scientific discovery. This halted any new knowledge of the heart until the Renaissance and the likes of Leonardo da Vinci and William Harvey.

Chapter Two

HEART AND SOUL

AS ANCIENT CULTURES became more advanced and had more time for contemplation, they began to question where in their body their mental abilities— their consciousness and reasoning— resided. Where was their incorporeal living essence, their soul? Some ancients believed the heart housed the soul—the cardiocentrists (from the Greek *kardia* or heart). Others thought the brain contained the soul—the cerebrocentrists (from the Latin *cerebrum* or brain). Most early cultures, including the ancient Sumerians, Egyptians, Chinese, Indians, some Greeks (namely Aristotle), and Romans were cardiocentrists, believing that the heart, not the brain, was the location of emotions, thoughts, and intelligence within the body.

Ptah was the ancient Egyptian creator god. Ancient Egyptians believed he existed before all things and used his heart to bring the world into existence. The Nubian pharaoh Shabaka had an inscribed stone made c. 700 BCE and claimed it was a copy of an earlier theological papyrus called the Memphite Theology from c. 2400–3000 BCE from the Great Temple of Ptah in Memphis. On the Shabaka Stone is this proclamation: "Ptah conceives the world by the thought of his heart and gives life through the magic of his word."

The ancient Egyptian word for heart was *ib*. Ib could mean the physical heart, the mind, intelligence, will, desire, mood, or understanding. Ancient Egyptians believed the heart was formed by one drop of blood from the mother's heart at conception and survived the physical death of the body. Upon death the ib was weighed against the Feather of Maat to determine whether a person led a virtuous life with Osiris escorting them to the Field of Reeds. If the ib weighed more than the feather, it was to be eaten by the goddess Ammit and the soul vanished from existence. Numerous expressions in the Egyptian language incorporated the word *ib*: *aA-ib* proud or arrogant (great of heart), *awt-ib* happiness (long heart), *aq-ib* intimate friend (trusted heart), *awnt-ib* greedy (covetous of heart), *bgAs-ib* to be troubled in the mind, *arq-HAty-ib* wise (perspicacious of heart), *dSr-ib* furious (make red the heart), *rdi-ib* devote (give heart to), and *ibib* favorite or love (heart to heart).[1]

The brain is not mentioned to any great extent in the medical papyruses from ancient Egypt. It was described only as an organ producing mucous, which drained out through the nose. Life and death were matters of the heart.

Egyptians were not the only ancient ones elevating the heart to soul status. Around the same time in history, the ancient Chinese came to believe that the heart was the "emperor" of the body and housed the mind.

■ ■ ■

Ancient Chinese believed the heart was the seat of consciousness, intelligence, and feelings. The heart held the *shen*, or spirit,
of a person. The heart had special importance in traditional
Chinese medicine. The heart was thought to be the "ruler" of all
the other organs, and when the body was healthy and balanced,
it was a kind and benevolent leader.

In Chinese philosophy, *xin* can refer to one's disposition or
feelings or to one's confidence or trust in something or someone. Literally, xin refers to the physical heart, although it is
sometimes translated as "mind." The ancient Chinese believed
the heart was the seat of the soul, of thought, intelligence, and
feelings.[2] For this reason, xin is also sometimes translated as
"heart-mind." In contrast, the brain was not included in the
organs of traditional Chinese medicine. What we now consider brain functions and diseases were the result of balanced
interaction between the *zang* organs: heart, liver, spleen, lung,
and kidney.

■ ■ ■

Like the ancient Chinese xin, the word for heart in Sanskrit
(the ancient language of Hinduism) was *Hridaya* or *Hridayam*,
which could also be translated as "seat of consciousness" or soul.[3]
It has been proposed that hridayam is made up of *hri*, which
means to receive; *da*, which means to give; and *ya* from *yam*,
which means to move around. Is hridayam a 3,500-year-old metaphor for the phases of the heart's contraction and circulation?

In ancient India, Shiva, the god protector, the destroyer of evil, and the regenerator of the universe and life, was also known as *Hridayanath*, "lord of the heart." His wife Parvati, the mother goddess, was known as *Hridayeswari*, "goddess of the heart."

■ ■ ■

Ancient societies universally taught that one's conscious being (the soul) resided in the heart. The Greeks probably knew of these teachings. When Western thought began to be developed in Greece, the location of the soul became a competition between Greek cardiocentrists and cerebrocentrists.[4] And the cardiocentrists had Aristotle as their quarterback, who said in c. 330 BCE: "The heart is the perfection of the whole organism. Therefore, the principles of the power of perception and the soul's ability to nourish itself must lie in the heart."

Several early Greek thinkers concluded that the soul was seated in the brain. The earliest may have been Alcmaeon of Croton (c. 500 BCE), who also believed semen was made in the brain and transported down the spinal cord. The most famous Greek cerebrocentrist was Hippocrates (c. 400 BCE), who based his theory mostly on Alcmaeon's work. But the Stoics and Aristotle and Praxagoras of Cos believed the heart was the most important organ of the body. Aristotle developed this belief during his observations of chick embryos because he saw that the heart was the first organ to form. "The heart is the place where life fails last of all: and we find universally that what is the last to be formed is the first to fail, and the first to be formed is the last to fail," he wrote in his *History of Animals* in 350 BCE.

Aristotle also wrote in *On the Parts of Animals* that the heart was "central, mobile, and hot, and well supplied with structures which served to communicate between it and the rest of the body." It was the central organ, "the source of all movement, since the heart links the soul with the organs of life." It therefore made sense to him that it was the location of the soul. The brain was far from the center of the body and was cold. The heart was warm, and warmth was equated with life. Aristotle believed the heart was the source of one's consciousness and intelligence. To Aristotle, the brain acted as a cooling unit, tempering the blood and heart with phlegm. This idea stuck, and the label *pituitary gland* in the brain derives from *pituita*, Latin for phlegm.

It isn't absurd that Aristotle thought that the heart was the seat of one's consciousness and soul. In fact, sudden strong emotions can cause an increased pulse, more forceful feeling heart beats, abnormal heart rhythms, heart attacks, and sudden death. We now know these responses are part of the heart-brain connection, but Aristotle observed this as a scientist. To him it made sense that the heart was therefore the seat of the soul. Aristotle's elevation of the heart over the brain as the center of one's being led to the heart maintaining this central position for almost two thousand years.

■ ■ ■

Claudius Galenus of Pergamum (129–216 CE), a Greek physician known as Galen who had moved to Rome in 162 CE for money and fame, agreed with Aristotle that the heart was the source of the body's heat, boiling the blood so that it turned from purple to red as it heated up.[5] Galen was considered to be the most important physician of the ancient world after Hippocrates. In *On the Usefulness of the Parts of the Body* (c. 170 CE), Galen wrote: "The heart is, as it were, the hearthstone and source of the body's innate heat and as the organ most closely related to the soul." However, he disagreed with Aristotle that the brain cooled the heart, arguing that it would need to be closer to the heart to do that.

After studying the teachings of Hippocrates and Plato, Galen came to believe in a tripartite soul structure. He used Platonic terms, describing the rational soul in the brain, the spiritual soul in the heart, and the appetitive soul in the liver. The brain directed cognition; the heart emotions.

Most early cultures were cardiocentric, believing the heart and not the brain was the location of the mind and soul within the body. These beliefs held sway in China (traditional Chinese medicine) and India (Ayurvedic medicine) for centuries to come. In the West, although there were some cerebrocentrists such as Hippocrates and Plato, the teachings of Aristotle and Galen were taken as gospel by the Catholic Church. These beliefs were accepted—to believe otherwise was sacrilege—for the next 1,500 years throughout the Dark Ages of Europe.

Chapter Three

THE HEART AND GOD

ANCIENT CIVILIZATIONS conceived of gods or a single God to explain their existence and the creation of the universe. Most cultures believed God was inside each person; in their heart. For many, the way to connect with God was through their heart.

In ancient India, the Upanishads were religious and philosophical Sanskrit treatises contained in the four Vedas, written late in the Vedic era of India (1700 to 400 BCE). They played an important role in the development of spiritual ideas in ancient India. They described the heart as the location of *Brahman*, Vedic Sanskrit for the cosmic or universal principle, or God. The Chandogya Upanishad states: "This is my Soul in the innermost heart, greater than the earth, greater than the mid-region, greater than heaven. This Soul, this self of mine, is that of Brahman."

The heart held the soul in ancient India and was responsible for all thoughts and emotions.[1] It was the location of one's self. The heart was the link between heaven and earth where one experienced the love of Brahman. The heart was where the soul rested and was the seat of divine love. The Brihadaranyaka Upanishad declared: "The heart, O Emperor, is the abode of all things, and the heart, O Emperor, is the support of all beings. On the heart, O Emperor, all beings rest. The heart truly, O Emperor, is the supreme Brahman. His heart does not desert him, who, knowing thus, worships it."

■ ■ ■

Confucianism, one of the most influential religious philosophies in history, was developed by Master Kong (Confucius) in the sixth to fifth centuries BCE. The main goal of Confucianism was the attainment of inner harmony with nature. Confucius taught that "wheresoever you go, go with all your heart." He believed that the heart, when unhindered by the schemes of the head, morally guided us: "If you look into your own heart, and you find nothing wrong there, what is there to worry about? What is there to fear?"

Master Meng (Mengzi), a fourth-century BCE Chinese thinker whose importance in Confucianism is second only to that of Confucius himself, taught that learning is to be the business of "seeking the lost heart . . . looking for it, you will get it; neglecting it, you will lose it."

■ ■ ■

In Buddhism, another of the Eastern ethical religions, the *Heart Sutra* is one of the most commonly recited and studied scriptures.

It is one of forty sutras that comprise the *Prajnaparamita Sutras*, which were believed to have been written between 100 BCE and 500 CE. Buddhists recite the Heart Sutra during their daily meditation. The *Prajnaparamitahrdaya* (The Heart of Perfection of Wisdom), or *Heart Sutra*, describes the nature of emptiness, or *sunyata*, a principal concept in Buddhism.

Zibo Zhenke (1543–1603 CE), one of the four great Buddhist masters of the late Ming Dynasty, wrote of the Heart Sutra, "This sutra is the principal thread that runs through the entire Buddhist Tripitaka. Although a person's body includes many organs and bones, the heart is the most important."

■ ■ ■

For the monotheistic Abrahamic religions further west, people connected with God through their heart. The word *lev*, or heart, appears in the Hebrew Torah (700–300 BCE) more than seven hundred times. For the Hebrews, the heart was the location of God's presence in the individual. The heart was one's center for spiritual, moral, emotional, and intellectual actions.

I will give them a heart to know me, that I am the Lord. (Jeremiah 24:7)

Do not look on his appearance or on the height of his stature, because I have rejected him. For the Lord sees not as man sees: man looks on the outward appearance, but the Lord looks on the heart. (1 Samuel 16:7)

Above all else, guard your heart, for everything you do flows from it. (Proverbs 4:23)

But the heart could be a source of good or evil.

The law of God is in his heart; none of his steps shall slide. (Psalms 37:31)

A wise person's heart leads the right way. The heart of a fool leads the wrong way. (Ecclesiastes 10:2)

■ ■ ■

The heart appears another 105 times in the Christian New Testament (50–150 CE). The heart held the knowledge of God within its walls. The heart allowed one to achieve the higher love of God. Early Christians believed that the heart was the seat of the soul. Not only was it the center of one's spiritual activity but of all the mental and physical operations of human life.

You yourselves are our letter of recommendation, written on our hearts, to be known and read by all. And you show that you are a letter from Christ delivered by us, written not with ink but with the Spirit of the living God, not on tablets of stone but on tablets of human hearts. (2 Corinthians 3:2–3)

They demonstrate that God's law is written in their hearts, for their own conscience and thoughts either accuse them or tell them they are doing right. (Romans 2:15)

Blessed are the pure in heart, for they will see God.
(Matthew 5:8)

To "love the Lord thy God with all thy heart" is repeated in the New Testament books of Matthew, Mark, and Luke.

St. Augustine of Hippo, in his Confessions (c. 400 CE), described the *cor inquietum* or "restless heart" as a heart divided between love of this world and love of God.[2] He wrote that every heart holds a divine spark. If lit, the heart flares into a holy glow and becomes one with God. The flaming heart became the icon of St. Augustine in the religious art that followed (figure 3.1).

In the twelfth century, the French abbot Bernard de Clairvaux, who was ultimately sainted, penned prayers to the *Cor Jesu Dulcissimum* ("the very sweet heart of Jesus"), which helped to establish one of the most well-known and widely practiced devotions of the Catholic Church: the Devotion of the Sacred Heart. The Sacred Heart, emitting rays of light, with a wound from an arrow, and sometimes wrapped in a crown of thorns, became the symbol of Jesus Christ and his love for humanity. This image of the heart had become an object of worship and became a common theme in medieval and Renaissance art (figure 3.2).

■ ■ ■

Physiological and anatomical knowledge of the heart, and even mention of heart diseases, can be found in Islamic sacred writings: the Quran (revelation given directly by Allah to

FIGURE 3.1 Portrait of St. Augustine of Hippo receiving the Most Sacred Heart of Jesus, by Philippe de Champaigne, seventeenth century.

Source: Los Angeles County Museum of Art / Wikimedia Commons / Public Domain.

FIGURE 3.2 Heart of Jesus in the context of the Five Wounds (the wounded heart here depicting Christ's wound inflicted by the Lance of Longinus), fifteenth-century manuscript (Cologne Mn Kn 28–1181 fol. 116).

Source: http://www.ceec.uni-koeln.de / Wikimedia Commons / Public Domain.

Muhammad; 600s CE) and the Hadith (practical lessons pronounced by Muhammad; 700–800s CE).

The heart appears 180 times in the Quran. The heart in early Islamic teachings was the center of feeling and reasoning.

Healthy hearts were pious and rational, whereas diseased hearts were inhumane and lost the capacity to see and understand.

> Thus, it is. And whosoever honors the Symbols of Allah, then it is truly from the piety of the heart. (Surah Al-Hajj, The Pilgrimage: 32).

> Have they not travelled through the land, and have they hearts wherewith to understand and ears wherewith to hear? Verily, it is not the eyes that grow blind, but it is the hearts which are in the breasts that grow blind. (Surah Al-Hajj, The Pilgrimage: 46).

Current medical knowledge of the heart, describing it as nothing more than a "blood pump," has demystified the heart. The heart is no longer the repository of the soul or the location of a relationship with God. However, the heart figuratively remains a symbol of devotional love; many still say, "Give your heart over to God." For the devoted, God does not reside in one's head but in one's heart.

Chapter Four

AN EMOTIONAL HEART

THE STORY OF RAMA (figure 4.1), as told in the ancient Indian Sanskrit epic *Ramayana* (600s BCE), places love and devotion in the heart.

After Lord Rama came back from his exile of fourteen years, having slain evil Ravana [a multiheaded demon king] and his demon followers, he was coronated the King of Ayodhya. In celebration, precious ornaments and gifts were distributed to everyone. Hanuman, Rama's general and ardent devotee, was gifted a beautiful necklace of pearls by Sita, wife of Rama.

Hanuman took the necklace, carefully examined each and every pearl, and threw them away. All were surprised by his behavior.

When asked as to why he was throwing away the precious pearls, he replied that he was looking for Rama in them. Thus, they carried no worth to him since anything in which there is no Rama is without worth.

FIGURE 4.1 Rama and Sita appearing on the heart of Hanuman.
Source: Karunakar Rayker / Wikimedia Commons / Public Domain.

When mockingly asked if Lord Rama was in Hanuman himself, he tore his chest apart to reveal his heart. The onlookers, now convinced of his genuine devotion, saw the image of both Rama and Sita appearing on his heart.

Ayurvedic practitioners of ancient India believed the human heart was actually two: the physical one that transported nutrients to the body and the emotional one that experienced love, desire, and sorrow.[1]

The *Sushruta Samhita* (sixth century BCE), contained in the four Vedas, states that the emotional heart, or heart's desire, began in the womb:

In the fourth month the several organs become more distinct, and as the heart of the fetus is already formed there, the vital functions begin to manifest themselves; for the heart is the seat of the vital functions; therefore, in the fourth month of pregnancy the fetus begins to exhibit a desire for various objects of sense, and we may notice the phenomenon of what is called the "longing."

Therefore, a pregnant woman when she, as it were possesses two hearts, may be called a "Longing Woman" and her longings should be satisfied; for, if they are not satisfied the child is apt to be humped, handless, lame, dull, short, squinted, sore-eyed, or totally blind. So, whatever a pregnant woman might desire should be given to her. When her longings are thus satisfied she begets a son who is valiant, strong and of prolonged life.

When I asked my mother if all her longings were satisfied when she was pregnant with me, she shrugged and said, "son, I love you with all my heart . . . anyway."

■ ■ ■

Ancient Greeks, even many of the cerebrocentrists, still believed emotions were contained in the heart.[2] In the 700s BCE, Homer wrote in the *Iliad*, "Hateful to me as the gates of Hades is that man who hides one thing in his heart and speaks another." And Heraclitus wrote in the 500s BCE, "It is hard to contend against one's heart's desire; for whatever it wishes to have it buys at the cost of the soul." The heart was central to love, to courage and to life itself, not only for humans, but for the gods as well.

The god Apollo found Eros (god of love; Roman Cupid) working with his bows and arrows. Apollo told Eros to leave such weapons to mighty war gods like him. Infuriated, Eros climbed Mount Parnassus and unleashed two arrows. The first, sharp and gold-tipped, pierced Apollo's heart and he fell in love with Daphne, the beautiful nymph daughter of the river god Peneus. The other arrow, dull and lead-tipped, pierced Daphne's heart, giving her an intense aversion to love in her heart. Apollo doggedly pursued Daphne. To escape she pleaded for her father's help. To prevent Apollo from capturing her, Peneus transformed her into a fragrant tree, the laurel (*daphne* in Greek).

Greeks of Homer's era (twelfth to eighth centuries BCE) tried to solve the heart-brain dichotomy by teaching that there were two souls in the body: the *psyche* that was the eternal life soul, and the *thymos* that controlled emotions, urges, and desires. Homer believed the psyche was located in the head, and the thymos was in the heart. In addition to anger and desire, the heart was the source of courage and bravery. In Homer's *Iliad*, when Ajax rebukes Achilles, Achilles answers: "My heart swells with anger."

For the early Greeks the heart was identified with love. The poet Sappho lived on the island of Lesbos in the seventh century BCE. Surrounded by her female disciples, she wrote passionate poems like this one: "Love shook my heart, Like the wind on the mountain, Troubling the oak trees."

■ ■ ■

Plato, a student of Socrates, was more a philosopher than a scientist. In the *Republic* (c. 376 BCE), Plato wrote: "And those whose hearts are fixed on reality itself deserve the title of philosopher." Plato believed that humans were made by a divine creator who placed an immortal soul and two mortal souls in each person. Plato believed the head governed the body; he was a cerebrocentrist.

In Plato's *Timaeus*, he stated that the immortal soul was the ruler of the body, with two other inferior mortal souls located in the heart and the stomach. The hot, pulsating heart ruled over anger and pride, as well as one's sorrow. It was the source of erotic desire, whereas the brain was the source of true love. The part of the mortal soul that directed hunger and bodily functions was in the stomach. It's fascinating that in the fourth century BCE an ancient Greek is describing what many accept today to be true: the brain is responsible for our reasoning and consciousness, and the heart is the repository of our emotions. Plato the "scientist" described the heart as a "knot of veins and the fountain of the blood." He wrote that the lungs cooled the heart so the spirit was better able follow reason over hot emotion.

The ancient Romans adopted Plato's tripartite soul theory, based primarily on the teachings of Galen.[3] For centuries, these teachings spread to other societies, and the heart remained the home of emotions. Even if the heart was not considered home to the immortal soul, it endured as the place where love and desire, anger and sorrow, resided in the body. This belief remained unchanged for another fifteen hundred years throughout the East and the West.

Pliny the Elder, a Roman military commander and author of the first encyclopedia, *Naturalis Historia*, died in 79 CE. Perhaps most famous for writing "Home is where the heart is," he died trying to rescue a friend and his family by ship during the eruption of Mt. Vesuvius. Like others during his time, he believed that the heart held one's love of home and family, and that people always carried home in their own being, in their heart.

Chapter Five

ANCIENT UNDERSTANDING
OF THE PHYSICAL HEART

The heart is an exceedingly strong muscle.
Hippocrates, 400s BCE

THE ANCIENT GREEKS began practicing medicine around 700 BCE.[1] Prior to this time and influenced by ancient Egyptian beliefs, these cultures thought illnesses were divine punishments from the gods. Alcmaeon of Croton (600 BCE), a student of Pythagoras, was one of the earliest Greek writers on the subject of medicine. He may be the first Greek to have performed anatomical studies on human bodies, both dead and living. Based on his experimental observations, Alcmaeon believed the brain was the location in the body of the mind and thought, the seat of sensations.

Like Alcmaeon, most ancient Greeks were cerebrocentrists (Aristotle being the major exception). For example, a person could be knocked unconscious (psyche in the head would be affected) while the body still lived (thymos in the heart kept body functions going). This incorrectly led many ancient Greek writers to believe that the head, not the heart, was the starting point for blood vessels. These vessels carried *pneuma* (life force) to the rest of the body, including the heart.

Ancient Greek physicians believed the heart was a furnace. Did the body not cool when the heart stopped beating? The heart, fueled by pneuma and blood from the brain and fanned by the breath, generated the body's heat.

■ ■ ■

Hippocrates of Kos is often referred to as the father of medicine. Doctors take the Hippocratic Oath when graduating from medical school as a promise to "do no harm." Hippocrates founded a medical school, and he was the first physician to teach that nature, not the gods, caused diseases. He established medicine as a discipline distinct from religion and philosophy.

The *Hippocratic Opus* is a collection of nearly sixty ancient Greek medical works associated with Hippocrates and his teachings (400s BCE to 100s CE). One text in the collection, *On the Heart*, recorded for the first time the anatomical details of the heart. According to Hippocrates's teachings, the heart was shaped like a pyramid and was a deep crimson color. It was contained in a membranous sac, which is known today as the pericardial sac. The sac was lubricated by fluid that helped absorb the heart's heat (pericardial fluid; think engine oil or brake fluid).

Hippocrates taught that if one removed the "ears" of the heart (these are the atria) the orifices of the chambers (the ventricles) were exposed. The ears worked the same way as the bellows of a blacksmith. Air was drawn in and pushed out with the expansion and deflation of the ears. The evidence for this purpose of the ears was the fact that when the heart pulsed the ears had a separate movement as the chambers inflated and collapsed (that is, the atria contract as the ventricles expand), this is called atrial-ventricular synchronization. *On the Heart* also pointed out that heart valves allowed for flow in only one direction. Hippocrates described the heart valve as a "masterpiece of Nature's craftsmanship."

One of the earliest descriptions of how to diagnose heart failure appears in the *Hippocratic Corpus*. The book describes a physical examination of the failing heart and fluid filled lungs this way: "When the ear is held to the chest, and one listens for some time, it may be heard to see the inside like the boiling of vinegar." Hippocrates may also be credited with the first description of sudden cardiac death when he said, "Those who suffer from frequent and strong faints without any manifest cause die suddenly." This death is due to loss of heart function, usually from a dangerous heart arrhythmia; it is the number one cause of natural death in the United States.

Based on Alcmaeon's work, Hippocrates believed that the brain, not the heart, was the seat of intelligence. However, Aristotle won this dispute, and most of Western civilization has believed that the heart was the home of the soul in the body right up to the modern age.

■ ■ ■

Aristotle was the first Greek to specifically describe chambers of the heart, although he observed three rather than four. He believed the right chamber (likely the right ventricle) had the most and the hottest blood.[2] The left chamber (likely the left atrium) had the least and the coldest blood. The middle chamber (likely the left ventricle) had a medium amount of the purest and thinnest blood. To Aristotle's credit, some argue he viewed the right atrium not as a heart chamber but as an engorged vein entering the heart.

We now know this is not correct. The method Aristotle used to kill animals for dissection may account for the differences in the amounts of blood he observed in each of the chambers. Aristotle strangled the animals, resulting in the veins and the right side of the heart being full of dark blood whereas the left side would be drained.

Aristotle correctly believed the heart was the center of a vascular system. He wrote in *On the Parts of Animals* (c. 350 BCE), "The system of blood vessels resembles the arrangement of watercourses in gardens which are constructed from a single source or spring [the heart] and divided up into many channels, each branching into many more, so that they supply water to every part." Aristotle believed that the brain served like a radiator to cool the hot heart. He believed more complex and rational animals produced more heat than simpler ones, such as insects. Thus, humans needed large brains to cool their hot, passionate hearts.

■ ■ ■

The Egyptian city of Alexandria was the center of Greek learning from the 300s BCE to the 600s CE. Alexander the Great founded the city in 331 BCE, which was ruled by the family of Ptolemy, one of Alexander's generals. Physicians performed human dissection in Alexandria as early as the 200s BCE. The government of the city even allowed vivisection, which is the gruesome dissection of people while they are still alive. It was usually performed on criminals as their punishment. Alexandria was a mix of Egyptian and Greek cultures, and the practice of human dissection was acceptable because the Egyptian practice of embalming required opening the body and removing organs.

Herophilus of Chalcedon (335–250 BCE) and Erasistratus of Ceos (c. 330–250 BCE) were two notable physicians practicing in Alexandria. They were responsible for first naming Hippocrates the "father of medicine." In his first-century medical treatise *De Medicina*, the Roman physician Aulus Cornelius Celsus (25 BCE–50 CE) wrote this about Herophilus and Erasistratus:

> Moreover, as pains, and also various kinds of diseases arise in the more internal parts, they hold that no one can apply remedies for these who is ignorant about the parts of them themselves; hence it becomes necessary to lay open the bodies of the dead and to scrutinize their viscera and intestines. They hold that Herophilus did this in the best way by far, when they laid open men while still alive—criminals received out of prison from the king—and whilst they were still breathing, observed parts which beforehand nature had concealed.[3]

Herophilus is considered the father of anatomy and physiology, and he believed the study of medicine required understanding the human body through human dissection. Herophilus was one of the first to conduct public anatomical dissections. These resulted in his discovery of the nervous system, which led him to believe that the brain, not the heart, was the thinking organ; he was a cerebrocentrist.

Herophilus was also the first to describe differences between arteries and veins. He noted that a corpse's veins would collapse when drained of blood, but the muscular arteries retained their suppleness. However, Herophilus incorrectly believed that dilation of the arteries drew in pneuma (air or spirit) from the heart and contraction of the arteries moved the pneuma and some blood forward, causing the pulse. Erasistratus, a disciple and collaborator of Herophilus, came very close to understanding circulation. He theorized that connections between arteries and veins had to exist but were too small to see, anticipating William Harvey's discovery of circulation by 1,800 years.

Unlike Aristotle, Erasistratus also believed that the brain, not the heart, was the directing organ of the body. He was the first to suggest that the heart was not the seat of the soul but just an organ responsible for heating the body.

Erasistratus lived for a time at the court of the Syrian king Seleucas (358–281 BCE). The king's son, Antioches, had fallen ill and was wasting away. Upon examining the prince, Erasistratus could find nothing wrong. But one day he noted that when the prince's stepmother, Stratonice, was near the young man's pulse began to beat fast and his skin grew hot. Erasistratus wearily informed the king of his diagnosis. The wise seventy-year-old king separated from his wife and married her to his son, who was then cured of his "sick heart."

■ ■ ■

The Greeks established anatomy as a legitimate way to study the workings of the heart and body. These anatomical studies continued with the rise of the Roman Empire, although most were done by Greeks who emigrated there. The most famous, Galen, wrote in the second century in his *On the Usefulness of the Parts of the Body*: "The heart is a hard flesh, not easily injured. . . . In hardness, tension, general strength, and resistance to injury, the fibers of the heart far surpass all others. For no other organ performs so continuously, or moves with such force as the heart."[4] But the question remained, did the head or the heart house emotions, memory, and thought?

The ancient Romans believed that the heart sustained life and held one's love. The Roman author Ovid (43 BCE–c. 17 CE) wrote: "Although Aesculapius himself applies the herbs, by no means can he cure a wound of the heart." Aesculapius was the Roman god of medicine, and it is his staff with a snake curled around it that is used as a symbol of medicine today. Venus, the Roman god of love, with the help of her son Cupid, targeted lovers' hearts with his arrows.

Ancient Romans believed diseases were inflicted as punishment by the gods.[5] Doctors were shunned. Unlike Greek physicians, Roman doctors were forbidden to dissect corpses, so they were limited in their understanding of the heart and body.

The Romans conquered Alexandria in 30 BCE, resulting in the suicides of Cleopatra and Mark Antony, the rulers of Egypt. The Romans became aware of the Greeks' medical knowledge and research on anatomy; primarily the works of

Herophilus and Erasistratus. The Romans captured physicians and brought them back to Rome, first as prisoners of war and later of their own volition so they could earn money in Rome.

The Romans soon began to adopt Greek medical and scientific ideas, but they did little to advance theories on the heart and the vascular system until the Greek, Galen, moved to Rome in 162 CE. His theories on the heart and body were later adopted by the Catholic Church as doctrine, making him the most important figure in Western medicine from the third to the seventeenth centuries (yes, for 1,500 years).

Galen had spent time in Alexandria dissecting animals and humans, and he became an expert in anatomy. He was allowed to dissect hanged criminals. He also was a doctor to the gladiators, and one can imagine the too real anatomy lessons he experienced from their open wounds as they lay dying. Galen wrote that gladiator wounds were "windows into the body." Galen quickly became a celebrity in Rome. He would perform dissections and treat patients in public for money. Because of this growing fame, Emperor Marcus Aurelius chose him as his personal physician.

Galen critically read the prior Greek medical writings. He experimented to prove or disprove theories on the heart and blood vessels. He especially appreciated the works of Herophilus and Erasistratus, although he loved to correct them. In *On the Usefulness of the Parts of the Body* (c. 170 CE), Galen observed: "Even if it would appear to be like muscles, it is clearly different from them. For muscles have fibers going only in one direction. . . . But the heart has both length-wise and cross fibers, as well as a third kind running diagonally, inclined at an angle."[6]

This important finding by Galen is now an intense area of research in modern heart science. Most understand that the heart (the left ventricle) expands and contracts in a concentric manner to eject blood out to the body. Envision a balloon inflating and deflating. But the heart also twists as it contracts to maximize the pump function. Think of twisting a wet dish towel as opposed to just squeezing it between your hands to get the excess water out. These heart muscle fibers, oriented in three different directions, optimize the heart's contraction, squeezing from the sides, the top, and the bottom and twisting in concert.

We now know much of what Galen theorized about the heart was incorrect. He believed that the muscular septum between the two ventricles had tiny holes in it to pass blood between ventricles. He was also incorrect that arteries carried pneuma (air mixed with some blood creating the vital spirit in us). Galen argued that the heart was secondary to the liver in its importance to the operations of the body. Galen incorrectly believed that digested food reached the liver from the intestines, where it was then converted into blood. He theorized that the liver, not the heart, was the source of blood vessels. The blood, after reaching the heart, was carried to all parts of the body and was converted into flesh.

Galen, like his predecessors (such as Aristotle), believed the main function of the left ventricle was to create heat for the rest of the body, and he compared the left ventricle to a coal furnace (the heart was not viewed as a pump because pumps did not yet exist). Inspired air was to cool the innate heat of the heart. However, Galen did fairly accurately describe how the heart works. He observed that the coronary system of vessels provides the blood supply to the heart itself. Through his animal dissections, he also correctly noted that not all animals have the same number of heart chambers. Fish, for example, have just one ventricle.

Galen experimented before he would accept Erasistratus's theory of a connection between the arterial and venous trees (now known as capillaries). He proved this connection by killing animals and severing their large arteries to "bleed" them out. This resulted in empty veins, along with empty arteries, confirming their connection. He also observed that almost all arteries were accompanied by a vein next to them, so it was likely that they were connected. Unfortunately, Galen did not make the next logical step in discovering the circulatory system. He continued to believe what Praxagoras had taught Herophilus and Erasistratus: the arteries mostly carried pneuma (air) to the rest of the body.

Although one might wonder how he figured this out—and shudder—Galen observed that fetal blood takes up inspired air from the mother's blood in the placenta and that this air-containing blood circumvents the fetal right heart and lungs (not yet functioning in the unborn child) and proceeds directly to the left heart through a hole between the atria, and then moves on to the arteries of the fetus. He further observed that after birth this hole closes and the path of blood flow switches, now going through the right heart and lungs before arriving at the left heart in the newborn.

In *On the Affected Parts*, Galen wrote, "There are three main factors which govern the body. It has been shown that, beside the heart as principle [organ], the brain is the foremost source of sensitivity and motility for all parts of the body, whereas the liver is the principle of the nutritive faculty. Death results from an imbalance of the humours in the heart, since all parts [of the body] deteriorate simultaneously with the heart."[7] Despite his great admiration for Aristotle, Galen built on the Platonic doctrine of the tripartite soul. The brain was starting to take over some, but not all, of the heart's presumed functions. Motor and sensory function resided in the brain, but the heart remained the emotional soul.

Galen revolutionized the ancients' understanding of anatomy and physiology through his experiments on the human body, especially the heart and its valves, as well as the arterial and venous trees. But his mistaken theories were also perpetuated, among them were that the presence of pores in the heart septum connecting the right and left ventricles; that the liver produced blood from food and was the origin of the body's arteries; that the heart's and vessels' function were to distribute the spirit (pneuma) throughout the body; and that the emotional soul was located in the heart. Many of Galen's theories held sway all the way into the seventeenth century because most of Western civilization, and thus medical knowledge, fell into the Dark Ages after the fall of the Roman Empire in 476 CE. For the next 1,500 years, little progress was made that advanced the scientific understanding of the heart and the vascular system.

Chapter Six

ANCIENT HEART DISEASE

Although commonly assumed to be a modern disease, the presence of atherosclerosis in premodern human beings raises the possibility of a more basic predisposition to the disease.
Randall C. Thompson et al., 2013

WE THINK OF heart attacks as a modern disease. We now live longer, eat more, exercise less, get fat, become diabetic, and smoke. As a result, we develop atherosclerosis, or "hardening of the arteries," due to cholesterol plaque buildup on the inner lining of the vessels that feed the heart muscle, the coronary arteries. Certainly we can assume that our distant ancestors of five thousand years ago had a very different lifestyle and were not at risk for atherosclerosis.

Well, they were.

Pharaoh Merenptah of ancient Egypt, who died in 1203 BCE at about seventy years of age, was plagued by atherosclerosis.[1] He was one of twenty mummies from the Egyptian National Museum of Antiquities in Cairo that were studied with CT scans in 2009. In sixteen of the mummies studied, the arteries and heart could still be seen. Of these, nine had atherosclerosis (56 percent!).

A larger study of mummies from different civilizations around the world suggested that atherosclerosis was not uncommon in ancient humans.[2] This study performed whole body CT scans on 137 mummies, spanning more than four thousand years, from four geographic areas with different dietary habits: seventy-six were ancient Egyptians who ate high-fat diets; fifty-one were ancient Peruvians who ate corn and potatoes; five were Ancestral Puebloans of Southwest America who were forager-farmers; and five were Unangan people of the Aleutian Islands who were hunter-gatherers. CT scans found atherosclerosis in 34 percent of the mummies. In mummies estimated to be more than forty years old at the time of death (that was old age back then), half had atherosclerosis. The study authors suggested that atherosclerosis is "either a basic component of aging, or that we are missing something very important that is a cause of atherosclerosis." The authors speculated that frequent infections could have caused ancient humans to suffer from chronic inflammation. Chronic inflammation can cause cholesterol buildup in artery walls, resulting in atherosclerosis. Ancient peoples also tended to cook and warm by fires, likely frequently inhaling smoke.

The 5,300-year-old Copper Age Tyrolean Iceman was found preserved in 1991 on the Tisenjoch Pass, in the Italian part of the Alps. A study of his DNA showed he was at an increased risk for atherosclerotic heart disease.[3] The Iceman's DNA revealed several single nucleotide polymorphisms (variations in a single DNA building block) that are linked with atherosclerosis in modern humans. But the Iceman did not die of a heart attack. In fact, he was shot in the back with an arrow. None of his contemporaries were likely to die this way either. Ancient humans had much shorter life spans then we do today, so they were less likely to die of a heart attack in "old age." But it's in our DNA! With the growth of civilization, ancient peoples at the top of society's hierarchies ate more than they needed to sustain life, became lazier and fatter, and succumbed to heart attacks and heart failure. They just did not yet know that it was heart disease.

Ancient Egyptians described what we now recognize to be accounts of heart attacks (chest pains often followed by death), and ancient Greeks described heart failure (shortness of breath with frothy sputum and swollen legs followed closely by death). However, there is nothing written to suggest that ancient societies made the connection between these sufferings and the heart. This lack of association between symptoms and heart diseases would persist for another 1,500 years. Only when civilization came out of the Dark Ages into the light of the Renaissance was this connection recognized.

PART 2

THE HEART GOES INTO THE DARKNESS
AND COMES OUT IN THE LIGHT

Chapter Seven

THE DARK AGES

Many afflict themselves more outwardly but make less progress
before God who is the inspector of the heart rather than the work.
Peter Abelard, Medieval French philosopher
and theologian, c. 1140 CE

Now it is agreed that the soul, with respect to the act and power
of life is in the heart. It is therefore necessary that the heart be the
point of origin of all the nerves and veins through which the soul
accomplishes its operations in the members'.
Albertus Magnus or Saint Albert the Great,
de Animalibus, 1256 CE

THE MIDDLE AGES of Europe, or the Dark Ages, began in 476 CE
(fall of the Roman Empire) and ended in 1453 CE (fall of Con-
stantinople). During this time, the Catholic Church subjugated
all aspects of life, including beliefs about the body and health,
and scientific advancement in medicine and anatomy stopped.
Living conditions deteriorated, resulting in repeated epidemics
such as leprosy. The church preached that epidemics and illnesses were inflicted by an angry God for one's sins. Healing the
body was only achieved through healing the soul. A physician
could not help you; your only hope was a priest. This widespread
doctrine stopped any advancements in medicine or knowledge
of the heart and the body for a thousand years.

The Catholic Church considered the theories of Galen and Aristotle (neither were Christians) on the heart and body as the only acceptable truths on human anatomy and physiology.[1] Throughout the Middle Ages, Galen's writings were accepted doctrine, immune to scientific challenge. The church sought out and destroyed works from other scientists and physicians, and Greek and Roman knowledge of the heart and the body was lost to science during this time.

Aristotle had discovered the heart was the first organ to form in the embryo, and the church declared this must be where God placed one's soul. The soul resided in the heart until death, when it left through the mouth. Medieval Christians believed that God resided in one's heart and took notes on its inner walls. The heart was a muscular tablet. God recorded each generous or ungenerous act or thought in the heart, which were to be reviewed upon one's death.

God also communicated with believers through their hearts. As written in the Bible in 2 Corinthians 3:2–3: "the epistle of Christ ministered by us, written not with ink, but with the Spirit of the living God; not on tablets of stone, but on tablets of human hearts." For Christians in the Middle Ages, the eternal soul was housed in the heart. When one's heart stopped beating, the soul left the body and went either to heaven or to hell. The soul's destination was determined by what had been recorded on the walls of the heart during life. If you had heart pains in medieval times, your only hope (beyond salvation of your soul, which was accelerated by donations to the church) was through home remedies such as smearing oil made from rue and aloe on your chest, eating salted radishes while placed in a vapor bath, or eating cockles (bivalve mollusks with heart-shaped shells) cooked in milk.

The term *mind*, which had originated as the concept of memory, eventually came to overlap with the idea of the soul. Aristotle had believed that cognitive faculties of the mind, senses, and emotions were located in the heart. Galen eventually adopted neo-Platonic theories, situating the rational and immortal soul in the brain and the emotional soul in the heart. These views, thanks to the Church, held sway throughout Europe's Dark Ages.

■ ■ ■

Many stories circulated of saints whose hearts were cut open after their death only to find inside evidence of their love for God and Jesus.[2] Santa Chiara da Montefalco, also known as Saint Claire of the Cross, was an Augustinian nun who in 1294 fell into an ecstasy for several weeks. She had a vision of a fatigued Jesus carrying his cross, and she reached to help him. Jesus told her, "I have found someone to whom I can trust mine cross," and he implanted it in her heart. Upon her death, her heart was removed by four nuns who found within it a cross and a scourge (whip). This was reported in Battista Piergilius's *The Life of Sister Chiara of Montefalco* (1663): "They knew well enough that the heart is concave and divided into two parts, being whole only in its circumference; then Sister Francesca felt with her finger that in the middle of one section there ran a nerve; and when she drew it out, they saw to their amazement that it was a cross, formed of flesh, which had been ensconced in a cavity of the same shape as the cross. Upon seeing this, Sister Margarita began shouting, 'A miracle, a miracle.' "[3]

Confirmed by the regional bishop and a jurist panel, Claire was eventually canonized a saint in 1881. We now know that the inner walls of the ventricles are not smooth but have irregular muscular columns called *trabeculae carneae*, which can take many shapes.

In eleventh-century Christian theology, the image of the heart came to represent the heart of Jesus. The Sacred Heart, a flaming heart pierced by a lance, surrounded by a crown of thorns with a cross on top, became the symbol for Jesus Christ. Representing his love for humanity, the Sacred Heart was frequently seen in religious art of the Middle Ages.

Culturally, the heart began to take on new meanings. The heart became a symbol of sincerity, truth, fealty, and loyalty and was seen on the shields of crusaders and families' coats of arms (figure 7.1). The heart stood for love of family or love of God, and it became one of the most popular symbols in medieval heraldry (figure 7.2).

An odd funerary practice became popular among the European upper classes (especially British and French royalty) in the eleventh and twelfth centuries. The ritual evolved from the belief that the heart was one's spiritual and moral center.[4] The heart would be removed from the deceased and buried separately from the body at a place of worship. We now refer to this custom as postmortem ablation of the heart. If a knight died far away from home, perhaps on a crusade, his heart would be sent home for burial. Kings and queens of the period often had their hearts entombed in one cathedral and their bodies in another.

FIGURE 7.1 Coat of arms of the Principality of Lüneburg, originating with William of Winchester, Lord of Lüneburg (1184–1213 CE), who married Helena, daughter of Valdemar I of Denmark and therefore adopted the "Danish tincture" to the arms of his father, Henry the Lion.

Source: Christer Sundin / Wikimedia Commons / Public Domain.

England's Richard I was nicknamed Lionheart. He earned this nickname because of his great feats in battle, and minstrels at the time sang that he had ripped out and eaten the heart of a lion to acquire its courage. Richard was mortally wounded

FIGURE 7.2 Knight's armory in the Grand Master's Palace in Valletta, Malta.

Source: Alexandros Michailidis / Shutterstock.

by a crossbow bolt during the siege of Chalus near Limoges, France, in 1199, and his heart was buried separately from his body. His dying request was that his organs be buried on location and that the rest of his body be interred at the Abbey of Fontevraud. But he wanted his heart to be embalmed and buried at the Notre Dame cathedral in Rouen.

Dervorgilla of Galloway (1210–1290) was a thirteenth-century Scottish noblewoman whose third son became King of Scotland.[5] Her husband was John Balliol of Barnard Castle. He was an adviser to English King Henry III, a joint protector of Alexander III of Scotland during his minority, and the bequestor of Balliol College at Oxford. When Balliol died, Dervorgilla had his heart removed from his body and embalmed. She carried it everywhere with her in an ivory and silver casket until she died. She was buried at the Cistercian Abbey of Dulce Cor, or Sweetheart Abbey, which she founded in her husband's memory. Dervorgilla was buried clutching her husband's heart to her chest.

King Louis IX of France (Saint Louis) died during the Second Crusade in 1270 when an epidemic of dysentery swept through his army at Tunis. Most of his viscera were buried on the spot. His skeleton, obtained by boiling the body, was returned to France. But his heart was sealed in an urn and placed in the cathedral of Monreale, Sicily. This practice of postmortem ablation of the heart continued among Scottish nobles until the seventeenth century, and French nobles until the eighteenth century. As late as 1849, Frederic Chopin, dying of tuberculosis in Paris, requested that his heart be returned to his homeland. His sister had his heart removed before his body was buried, preserved in a jar of cognac, and secreted to Poland to be buried at the Holy Cross Church in Warsaw.

■ ■ ■

Things began to change in Europe between 1000 and 1200. European monarchs became owners of more territory, their wealth grew, and their courts became centers of culture. Universities, such as those at Bologna, Oxford, and Paris, were established. Learning began again to take root.

Europeans traveling to the Middle East on crusades (1096 to 1291) brought back Arabic texts on medicine and anatomy. These texts explained discoveries Islamic physicians had made about the heart and the body during Europe's Dark Ages. Their work was largely based on prior Greek and Roman medical theories that they had saved and preserved. If not for the Islamic translations of Greek and Roman medical thought, such works as those of the father of medicine, Hippocrates, would have been lost to history.

Studying Aristotle's rediscovered works in the thirteenth century, thinkers of this period developed complex theories of the mind and the soul. Albertus Magnus, a German Catholic Dominican friar and philosopher, agreed with Aristotelian cardiocentrism, which seated the soul in the heart. This was congruent with Christian writings of the Middle Ages stating that the heart housed the passions of the soul. Most thinkers of the age accepted Galen's neo-Platonic theory that the rational soul resided in the brain and the lesser spirited soul was located in the heart. The heart was still the location in the body of understanding and feeling.

In contrast, Albert's student St. Thomas Aquinas (c. 1224–1274), who also studied the newly discovered works of Aristotle, did believe that the heart directed a person's movement.[6] Aquinas revised the Aristotelian cardiocentrist's view, stating that the soul was not in the heart but in the form of the body. The heart was the mover of the body, but the heart was moved by the soul. The soul was necessary for the heart to beat, and the heart was directed by the soul through its emotions.

Possibly taking their lead from Aquinas, after one thousand years of doctrine the Catholic Church updated its view on where the soul existed in the body. At the Council of Vienna in 1311, ordered to convene by Philip IV of France who had attacked Rome and killed Pope Boniface VIII (who tried to excommunicate him), Philip famously directed his "new" Pope, Clement V, to withdraw papal support for the Knights Templar. But the first decree of the council declared that the soul was no longer housed in the heart but in the whole body: "In order that all may know the truth of the faith in its purity and all error may be excluded, we define that anyone who presumes henceforth to assert defend or hold stubbornly that the rational or intellectual soul is not the form of the human body of itself and essentially, is to be considered a heretic." The heart was beginning to lose its hold as the home of the soul. From the twelfth century on, as centers of learning were founded throughout Europe and the fear of committing heresy by disagreeing with church doctrine waned, scientists began to place cognitive functions in the brain.

Coincident with Europe's Dark Ages, other cultures had developed their own ideas about the heart and its importance in life and death. Islamic physicians and scientists, influenced by the earlier teachings of the Greeks and Romans, advanced theories on heart anatomy and its role in the human body, both physically and metaphysically. Meanwhile the Vikings to the north were worshipping the "cold" heart, and Mesoamericans were sacrificing vast quantities of "hot" beating hearts to appease the gods.

Chapter Eight

THE ISLAMIC GOLDEN AGE

Your heart knows the way. Run in that direction.
Rumi, 1207–1273 CE

The heart is a
Thousand-stringed instrument
That can only be tuned with
Love.

Hafiz, 1320–1389 CE

AS EUROPE fell into the Dark Ages for a thousand years, producing no significant advances in anatomy or medicine, Islamic thinkers expanded on the theories of the ancient Greeks and Romans.[1] They copied ancient medical texts that were destroyed by the Catholic Church in Europe, including Hippocrates, the Greek physicians of Alexandria, and Galen. If not for Islamic scholars and physicians, knowledge of the heart and medicine before 400 CE might have been lost and the Renaissance in Europe would have begun with no past knowledge to build upon. Physicians and scientists of the Renaissance learned of ancient Greek and Roman medical knowledge by reading Arabic translations of the long-lost texts.

For early Islamists, the heart was the center of emotions, intentions, and knowledge. At the same time, physiological and anatomical knowledge of the heart and heart diseases were found in Islamic spiritual writings in the Quran (seventh century) and the Hadith (ninth century). Heart diseases were thought to be related to negative emotions (e.g., anger or fear) or spiritual failures (e.g., sinning and unbelief).

Islamic physicians studied the earlier works of the ancient Greeks and Romans, learning and challenging these theories about the heart. They introduced the concept of medical school hospitals, where both men and women practiced medicine.[2]

■ ■ ■

The Persian physician and philosopher Abu Bakr Muhammad ibn Zakariyya al-Razi (865–c. 925 CE), known in the West as Rhazes, wrote *The Diseases of Children*, the first text to distinguish pediatrics as a separate field of medicine.[3] He was the first to identify a fever as a defense mechanism against disease and infection.

Al-Razi was the first physician to use the term *sudden death* (more than one thousand years ago). He observed that the heart was responsible for syncope (loss of consciousness) and sudden death (death due to loss of heart function). We now know this is caused by dangerous heart arrhythmias, and it is the number one cause of natural death worldwide. Al-Razi wrote, "Sudden death takes place when the heart contracts but does not relax."

Al-Razi went on to explain that "in the heart, there are eight types of bad tempers: blockage in its arteries, blockage in its opening and swelling followed by irregular pulses, fast and then syncope."[4] In this quote Al-Razi describes what we now know as (1) atherosclerotic coronary artery disease, (2) heart valve stenosis, (3) heart failure, and (4) life-threatening heart arrhythmias. Yet in his book *Spiritual Medicine*, Al-Razi followed Plato's and Galen's concepts of tripartite souls: (1) the appetites (including sensual desire) were located in the liver; (2) the spirited, hot-blooded soul (think emotions) was located in the heart; and (3) the rational or divine soul was in the brain.

■ ■ ■

'Ali ibn al-'Abbas al-Majusi (925–994 CE), known as Haly Abbas in Europe, was court physician of Adud al-Dawla, king of Persia. He was the founder of the Adudi Hospital in Baghdad, where he wrote *The Complete Book of the Medical Art*. Al-Majusi rejected some of Aristotle's and Galen's theories. On the question of the arteriovenous system, Al-Majusi made distinctions between arteries and veins based on their thickness and function. He was one of the first to suggest a connection between the arterial and venous systems. In *The Complete Book of the Medical Art*, he wrote: "There are some foramina within the non-pulsating vessels [veins] that open to the pulsating vessels [arteries]." This predates the discovery of capillaries by almost seven hundred years.

■ ■ ■

Abū 'Alī al-Ḥusayn ibn 'Abd Allāh ibn al-Ḥasan ibn 'Alī ibn Sīnā, known as Ibn Sina (980–1037 CE) or as Avicenna by Europeans, was a Persian physician, astronomer, and philosopher. He became known in Europe as the "Prince of Physicians." His major work, the *Canon of Medicine*, completed in 1025, was the essential medical text for scholars of Islam and Europe for more than six centuries. In another major work, the *Book on Drugs for Heart Diseases*, Ibn Sina discussed therapies for breathing difficulties (possibly acute heart failure), palpitations, and sudden loss of consciousness (syncope). Ibn Sina was the first physician known to recommend regular exercise and a healthy diet to prevent heart disease!

Ibn Sina made great progress in the anatomical knowledge of the heart. He recognized the origin of the arteries coming from the left side of the heart (the aorta and its branches, referred to as the "great vessels"). He identified the difference between the thickness of the left (thick) and right (thin) ventricle walls. In addition, he described the timing difference between the contractions of the atria and ventricles (cardiac atrioventricular synchrony). Unfortunately, he also wrote that the quantity and consistency of chest hair correlated with the strength of one's heart.

In the *Canon of Medicine* Ibn Sina wrote: "The heart is the root of all faculties and gives the faculties of nutrition, life, apprehension, and movement to several other members." He believed that to govern the body, the soul acted through the intermediacy of the heart to direct the other organs and generate the body's heat.

Like Galen, he wrote that the heart produced an "innate heat," and like the ancient Chinese, he believed that this hot heart controlled and directed the other organs of the body. He did posit that five internal senses resided in the brain: common sense, imagery, imagination, estimation, and memory. These internal senses were directed by the incorporeal mind (self), a concept not unlike modern views on the soul.

■ ■ ■

Ala-al-Din abu al-Hassan Ali ibn Abi-Hazm al-Qarshi al-Dimashqi of Damascus (1213–1288), also known as Ibn al-Nafis, challenged some of Galen's and Ibn Sina's assumptions about the heart, particularly the belief that invisible pores in the heart septum allowed blood to pass between the left and right ventricles.

Through animal dissection—he did not like dissecting human corpses because it was contradictory to the teachings of the Quran—he proposed the existence of the pulmonary circulation. In his *Commentary on Anatomy in Ibn Sina's Canon* (1242; fol. 46 r), he wrote:

> The blood, after it has been refined in the right cavity, must be transmitted to the left cavity where the (vital) spirit is generated. But there is no passage between these cavities, for the substance of the heart is solid in this region [the heart's septum] and has neither a visible passage, as was thought by some persons, nor an invisible one which could have permitted the transmission of blood, as was alleged by Galen.

Ibn al-Nafis postulated that the coronary arteries supplied the heart muscle, contradicting Galen who believed the heart received nutrients from blood flowing inside the heart chambers. He also challenged Aristotle's theory that psychic faculties (cognition, sensation, imagination, and locomotion) came from the heart. He argued that the brain and nerves were cooler than the heart and arteries so the psychic faculties logically came from the brain.

■ ■ ■

While Europe languished in the Dark Ages, Islamic scholars and doctors built on the works of the ancient Greeks and Romans, making new discoveries and advancing human knowledge of the heart and the body. Europeans studied this medical knowledge as they came out of the Dark Ages, recognizing the importance of these contributions by Islamic scholars to the understanding of the workings of the heart and medicine. (Remember, they referred to Ibn Sina as the Prince of Physicians.)

In the "General Prologue" to the *Canterbury Tales* (Geoffrey Chaucer, c. 1400), one of the pilgrims (a physician) traveling from London to the Canterbury Cathedral credits Al-Razi (Rhazes), Al-Majusi (Haly Abbas), and Ibn Sina (Avicenna), along with Hippocrates and Galen, as the historic figures from whom he learned medicine.

Chapter Nine

THE VIKING COLD HJARTA

William came o'er the sea,
With bloody sword came he:
Cold heart and bloody hand
Now rule the English land.

Snorre Sturlason, Heimskringla, 1230 CE

THE VIKINGS played a major role in northern Europe during the Middle Ages. The Viking Age, from the eighth to eleventh centuries, was characterized by extensive migration and commercial pursuits—okay, and some plundering. *Heimskringla* (Sagas of the Norse Kings), written by Icelandic poet and historian Snorre Sturlason in 1230, is a collection of biographical stories about Norwegian and Swedish kings from the ninth to twelfth centuries.[1] The epigraph to this chapter refers to the Norman king William the Conqueror.

William had a "cold" heart, which was a compliment by the Vikings. The smaller and colder the heart, the braver the warrior. A coward's heart was large, warm, and trembled. The brave had a small, cold, and firm heart. In Old Norse, the word *hjarta* could mean heart as in the muscle or heart as in the seat of emotions. It could also mean brave or hearted, as in "hardhearted."

In the Norse heroic poem *Atlakvitha* (eleventh century), Gunnar and his brother Hogni are captured by Atli's men (supposedly Atilla and the Huns). Atli wants their hidden treasures and demands Gunnar reveal the treasure's location. Gunnar responds:

First the heart of Hogni shall ye lay in my hands.

Atli agrees and they bring him a heart on a platter.

Here have I the heart of timid Hjalli, unlike the heart of
bold Hogni, for it trembles still as it sits on the platter;
more by half did it tremble in the breast of the man.

They go back for bold Hogni's heart.

Then Hogni laughed when they cut out the heart of the
living helm-hammerer; tears he had not.

Then Gunnar spoke as they brought him his brother's heart:

Here have I the heart of bold Hogni, unlike the heart of
timid Hjalli, little it trembles as it lies on the platter,
still less did it tremble when it lay in his breast.

Gunnar then laughed at them, for with his brother dead he was now the only one who knew where the hoard of treasures was hidden. Despite torture he would not reveal the treasure location. Eventually they gave up and threw him into a pit of vipers, where he died as he played the harp.

The Vikings were in actuality farmers, and only part-time warriors (both the men and the women), but the courageous, hard, cold heart was their ideal. As Sigurd, a legendary Norse dragon slayer, states in the *Volsunga Saga* (thirteenth century), "Whenas men meet foes in fight, better is stout heart than sharp sword."

Beowulf is an Old English epic poem that tells the story of Beowulf, hero of the Geats. Beowulf comes to the aid of Hrothgar, king of the Danes. Hrothgar's mead hall, Heorot, has been attacked by the monster Grendel. The story takes place in sixth-century Scandinavia. In the earliest manuscript from 975 CE (though the story likely existed in oral form from as early as the seventh century), Beowulf states:

> when he comes to me
> I mean to stand, not run from his shooting
> Flames, stand till fate decides
> Which of us wins. My heart is firm,
> My hands calm: I need no hot
> Words.

By the end of the eleventh century, Viking kingdoms were becoming Christianized, including the Danish kingdom of Harald Bluetooth and the English kingdom of William the Conqueror. Thus, the Vikings eventually traded their cold heart for the Cor Jesu Dulcissimus, the very sweet heart of Jesus.

Chapter Ten

AMERICAN HEART SACRIFICE

DURING TOXCATL, the fifth month of the Aztec ritual calendar, a young man was chosen for a great honor based on his looks. He needed to have smooth skin and long, straight hair. For the next year, this man would be treated like a god—literally.

He would be dressed up as Tezcatlipoca, god of the summer sun and night sky, who could ripen the crops or send a burning drought to kill the plants. His skin would be painted black, and he would wear a flower crown, a seashell breastplate, and lots of jewelry. The man would be given four beautiful wives to do with as he pleased. He was only asked to walk through the town playing a flute and smelling flowers so that the people could honor him.

When twelve months had passed, this god impersonator walked up the stairs of a temple pyramid, breaking his flutes as he climbed to the top. As an adoring crowd watched, he laid on a stone altar. Four priests held his arms and legs as a fifth sliced open the young man's upper abdomen. Reaching in with his hand, the priest tore out his heart. Lifting the beating heart up to the sky in offering, the sun and rains would surely come to ripen the crops (figure 10.1).

FIGURE 10.1 Aztec ritual heart sacrifice. Codex Magliabechiano, Folio 70.
Source: Foundation for the Advancement of Mesoamerican Studies, Inc. / Wikimedia Commons / Public Domain.

Afterward, a new, lucky young man would be chosen to represent Tezcatlipoca for the next year.

The Aztecs lived in what is now central Mexico (1325 to 1521 CE), and they believed the human body served as the temporary location for three souls, each residing in a different part of the body.[1] This was strikingly similar to the tripartite soul paradigms of Plato, Galen, and Al Razi. *Tonalli* was in the head, providing the body with reason, vigor, and energy for growth and development. A person's tonalli could leave the body during dreams or ritual hallucinogenic experiences. *Teyolia*, in the heart, was the source of knowledge, wisdom, and memory. Unlike tonalli, one's teyolia could not leave the body while the person was alive. Teyolia was the immortal part of the human that transcended to the afterlife. *Ihiyotl*, in the liver, controlled passions, emotions, and desires.

Aztecs came to know *teotl* (god or divine energy) using their hearts.[2] Situated between the head and the liver, the heart was centrally located to take advantage of the head's reason and the liver's passion. This idea was not that different from Aristotle's reasoning for the centralized placement of the soul in the heart.

The Mayans of Central America (c. 1800 BCE to 1524 CE) believed that human beings were created to nourish and sustain the gods. Blood contained vital power, and the heart held teyolia, which could strengthen the gods. Sacrificing human hearts to the gods (*nextlaoaliztli* in the Nahuatl language) meant "the giving of what is right or proper." Before and during heart sacrifices, priests and community members gathered in the plaza below the temples and stabbed, pierced, and bled themselves as auto-sacrifices. Women pulled ropes through their tongues, and men pierced their penises as mini sacrifices of blood to the gods.

Human sacrifice was common in many parts of Mesoamerica. Archaeological evidence suggests that heart sacrifices were taking place as early as the time of the Olmecs (1200–400 BCE). Early Mesoamerican cultures, such as the Purépechas (150 BCE–1500s CE) and Toltecs (900–1200s CE), regularly performed heart sacrifices.

Heart sacrifices were nothing new when the Aztecs rose to prominence in the twelfth to fourteenth centuries.[3] The sun god, Huitzilopochtli, was waging a constant war against darkness, and if darkness won, the world would end. To keep the sun moving across the sky, to keep their crops and themselves alive, the Aztecs had to feed Huitzilopochtli with human blood and hearts. When the Aztecs sacrificed people to Huitzilopochtli, the victim would be placed on a sacrificial stone. The priest would cut into the upper abdomen and up through the diaphragm with an obsidian or flint blade. He would then grab the heart and tear it out with his hand. The priest would then hold it up to the sky still beating as an offering. The Aztec gathered enormous clay jars filled with hearts that would later be poured into cenotes (large sinkholes full of groundwater) to appease the gods and in gratitude for sun-drenched crops.

After the heart was torn from the body, the rest of the body would be pushed down the pyramid to the Coyolxauhqui stone, named after the moon goddess Coyolxauhqui. Butchering the rest of the body on this stone reenacted the story of Coyolxauhqui, mother of the war god Huitzilopochtli. Huitzilopochtli, angered at his mother for not being willing to move from the sacred Snake Mountain, decapitated and dismembered her, eating her heart, after which he led the Aztecs to their new home. The dismembered body parts of the sacrificial victim were given to the warrior responsible for his capture. He then gave them to important people as an offering, or used the pieces himself for ritual cannibalism. During the fourteenth century, it is estimated that more than 15,000 sacrifices took place yearly in Tenochtitlan, the capital of the Aztec empire.

Even more grisly, a recent excavation in Peru found that the Chimu empire (1000–1400s CE) cut out the beating hearts of more than 140 children, aged six to fourteen, in a single day after flooding rains.[4] They believed a mass ritual killing of children was necessary so that their hearts would appease the gods and stop the rains. Another recent excavation found another 132 children sacrificed for their hearts—it happened more than once.

Pedro de Alvarado (1485–1541) was the only known conquistador to take part in the conquests of the Aztecs, Maya, and Inca. In *The True History of the Conquest of New Spain* (1568), Bernal Diaz del Castillo wrote:

When Alvarado came to these villages he found that they had been deserted on that very day, and he saw in the *cues* [temples or pyramids] the bodies of men and boys who had been sacrificed, the walls and altars all splashed with blood, and the victims' hearts laid out before the idols. He also found the stones on which their breasts had been opened to tear out their hearts. Alvarado told us that the bodies were without arms or legs, and that some Indians had told him that these had been carried off to be eaten. Our soldiers were greatly shocked at such cruelty. I will say no more about these sacrifices, since we found them in every town we came to.[5]

We know little of the Mesoamericans' practice of medicine. They did use herbs to treat various maladies, but they believed illnesses were due to the displeasure of the gods.

Once the pre-Columbian cultures of the Americas were conquered by the Spanish, between 1521 and 1532, heart sacrifices were prohibited. The Indigenous peoples were evangelized into Catholicism. Their god was now a "loving" god represented by a burning heart that represented Jesus's love for them. They could practice this new religion of love while they were forced to labor as slaves for their conquerors. Was this the end of their world as predicted when they could no longer sacrifice hearts to Huitzilopochtli? At least it appeared so metaphorically.

■ ■ ■

At the other end of the Americas were the Gwich'in,[6] the northern most Indigenous nation in the Americas. They had been hunting caribou for twenty thousand years, and they call themselves "the caribou people" because of the deep spiritual connection they share with caribou. In their creation story, the Gwitch'in and caribou were originally one. As they became separate beings, human and caribou each carried a piece of the other's heart. As every caribou has a bit of the human heart

and every human has a bit of caribou heart, they were spiritu-
ally, physically, and mentally connected. They knew each other's
habits, respected one another, and helped each other survive; the
caribou provided food and clothing for humans, and humans
only took what they needed and protected the habitat of the
caribou. Fast forward to the modern day and this synergistic
relationship of hearts is now at risk because of the encroach-
ment of oil drilling on their lands.

Chapter Eleven

THE HEART RENAISSANCE

AS THE DARK AGES gave way to the Renaissance and the Age of Discovery (also known as the Age of Exploration), scientists and physicians began to question long-standing theories on the heart.[1] These were mainly Galen's views, with some of Aristotle's and Hippocrates's mixed in, mostly from Arab translations. But the heart remained the chief organ of the body, and home to the emotional soul.

In 1498 Leonardo da Vinci, a painter, scientist, and inventor, wrote in his *Notebooks*: "Tears come from the heart and not the brain." In 1535 Andrés Laguna de Segovia, a physician and botanist, wrote in his *Anatomica Methodus*: "If indeed from the heart alone rise anger or passion, fear, terror, and sadness; if from it alone spring shame, delight, and joy, why should I say more?" Sounding a lot like the ancient Chinese view of the heart, in 1621 Robert Burton, author of *The Anatomy of Melancholy*, wrote: "The [heart is the] seat and foundation of life, of heat, of spirito, of pulse and respiration, the sun of our body, the king and sole commander of it, the seat and organ of all passions and affectations."

■ ■ ■

Two major figures of the Renaissance—Leonardo da Vinci and Andreas Vesalius (known as the father of modern anatomy)— furthered understanding of heart anatomy and sketched what we now accept as the first accurate representations of the heart. The polymath, Leonardo di ser Piero da Vinci (1452–1519), was an expert in anatomy.[2] Working with Marcantonio della Torre, da Vinci studied and sketched tendons, muscles, bones, and organs. Della Torre was a professor of anatomy at the University of Padua. Della Torre had permission to dissect human corpses from hospitals. He was intending to publish a book with da Vinci, but he died prematurely of the plague in 1511. During their time together, da Vinci created more than 750 detailed anatomical drawings with notes of the human body. Like da Vinci, most Renaissance artists viewed dissection as useful training in the details of the body. There were three elements of the body that needed to be studied for painting: the arrangement of the bones, the distribution and arrangement of the muscles, and the overlying skin and fat.

Many artists at the time (e.g., Michelangelo) studied the bones, muscle, and skin, but da Vinci was unusual in the extent to which he also examined the rest of the body's insides. He was the first to draw four chambers of the heart accurately (figure 11.1). Because of detailed anatomical studies, the heart

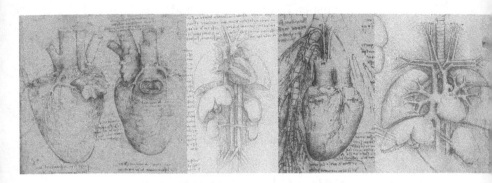

FIGURE 11.1 Heart drawings and annotations by Leonardo da Vinci.
Source: Collection Windsor Castle, United Kingdom. Royal Collection Trust/
© Her Majesty Queen Elizabeth II 2018.

was now correctly understood to be divided into four chambers. In 1535 Andrés Laguna de Segovia (1499–1559), a Spanish physician and pharmacologist, wrote:"The heart has only two ventricles, a right and a left. I do not know what is the meaning of the riddle proposed by the people who add a third ventricle to the heart unless perhaps they intend by it those pores which are found in the septum." He was obviously taking a shot at Galen.

Da Vinci established through experiments that Galen was wrong; blood, not air, entered the heart from the lungs. He also proved through experimentation that the valves allowed the blood to flow through the heart chambers in only one direction and prevented it from leaking backward. Da Vinci first described that the way the aortic valve closed was due to vortices, little eddies of blood, forcing the valve shut after blood had been ejected out of the left ventricle and through the aorta to the rest of the body. He figured this out by filling an ox heart with wax. Once the wax hardened, he re-created the aortic structure in glass. He then pumped water containing grass seeds through the glass aorta model and watched the seeds swirling back toward the valve. This observation was not proven again until 1968, when Oxford engineers Brian and Francis Bellhouse thought they were the first to discover this fact. It was not until a year after they published their work that they found da Vinci had beaten them to this conclusion four hundred years earlier.

Before we give da Vinci too much credit for his ability to draw the heart with a great deal of accuracy, it's important to understand that he did not deviate significantly from Galen's views. "The heart of itself is not the beginning of life but is a vessel made of dense muscle vivified and nourished by an artery and a vein as are the other muscles. The heart is of such density that fire can scarcely damage it." He believed the heart's main purpose was to create heat, which it did through the friction caused by movement of blood back and forth between its chambers.

Although da Vinci was heavily influenced by Galen's teachings, he did make some new discoveries that were the first real progress in our understanding of the heart in more than a thousand years. For example, it became clear to da Vinci that the heart—not the liver, as Galen believed—was the center of the arterial and venous systems. He also placed the soul in the brain—specifically above the optic chiasm in the anterior portion of the third ventricle—where it resided in the seat of judgment where all the senses came together. He called this the *senso commune*, or common sense.

Da Vinci was also the first to realize that narrowing and clogging of the coronary arteries (what we now know as coronary atherosclerosis) can cause sudden death. In 1506, he observed a man supposedly one hundred years old die suddenly and peacefully. Da Vinci performed "an anatomy to discern the cause of a death so sweet." His dissection led him to discover the "thickened coat," or narrowing in the old man's coronary arteries, deducing this as the cause of the man's sudden death. Thus, da Vinci may have been the first in history to diagnose coronary artery disease as a cause of sudden death.

No longer in the Dark Ages, da Vinci's new revelations on the structure and function of the heart were the first real progress in Western understanding in 1,500 years. However, after he died, all of his works passed on to his apprentice and friend Count Francesco Melzi. Melzi's descendants sold Leonardo's journals, and his work was lost or in the hands of private collectors. Eventually purchased by English King Charles II, da Vinci's anatomical notes and drawings of the heart ended up in the Royal Library at Windsor, and were forgotten. They were not rediscovered and published until 1796—more than 250 years after his death.

■ ■ ■

Andreas Vesalius, a Flemish anatomist and physician, left Belgium in 1533 to go to Padua in the kingdom of Venice,[3] which had become the center of Western science and medicine. But anatomy demonstrations were more like circus theaters than knowledge centers. Anatomists were entertaining crowds with knowledge they acquired from Galen and the Greeks. Vesalius wanted to challenge old theories, but he needed cadavers. Vesalius became one of history's most expert body snatchers. He cut down criminals from the gallows and removed half buried bodies from cemeteries. He and his students broke into ossuaries and stole bodies. The more he dissected, the more he questioned accepted theories on the heart and body.

One of Galen's theories that bothered Vesalius was that blood moved through invisible pores from the right to the left side of the heart. Vesalius studied hearts and did not see pores; instead he found a thick muscular wall separating the ventricles (the heart's septum). Unfortunately, he did not take the next step and discover the circulatory system. At the same time, he accepted some of Galen's erroneous theories, such as blood was produced by the liver and consumed in the body, and that the heart was a furnace.

Vesalius wrote what is considered one of the most important books in the history of medicine, *De Humani Corporis Fabrica* (On the Fabric of the Human Body). Published in 1543, Vesalius challenged much of what was known about the human body, correcting many of Galen's errors. Vesalius called the heart the "center of life." However, he avoided the subject of the soul's location, fearing to anger religious authorities who allowed only established church doctrine:

To avoid running afoul of some "idle talker" here, or some critic of doctrine, I shall completely avoid this dispute concerning the types of soul and their location. For, today, you will find many judges of our most truthful and sacred religion, especially among the natives of my country, and if they heard anyone muttering about the opinions of Plato, Aristotle, Galen, or their interpreters, about the soul, even when the subject is Anatomy (when these subjects are especially likely to be discussed), they leap to the conclusion that such a person is in doubt about the faith or has some uncertainty about the immortality of Souls.

Some believe Vesalius hired Titian (or an artist from Titian's school) to utilize art as a tool to illustrate the heart and organs in great detail (figure 11.2). He could diagram the paths of arteries and veins and their branches throughout the body. He also published the first drawings of valves in veins (they keep the blood going toward the heart and not backward to pool in the legs). Historians speculate that if Vesalius had not left Padua to become the personal physician of Charles V of Spain, it is probable that he would have discovered the circulation of blood to and from the heart.

In fact it was a student of Vesalius in Padua, Hieronymus Fabricius (1537–1619), who discovered the valves in the veins. He noticed that blood could not move from the heart toward the periphery through the veins. He correctly suggested that this was to prevent blood from pooling in the feet. But what he did not understand was that blood returned toward the heart through the veins. It was a student of Fabricius who figured that out. His name was William Harvey, and we will meet him shortly.

■ ■ ■

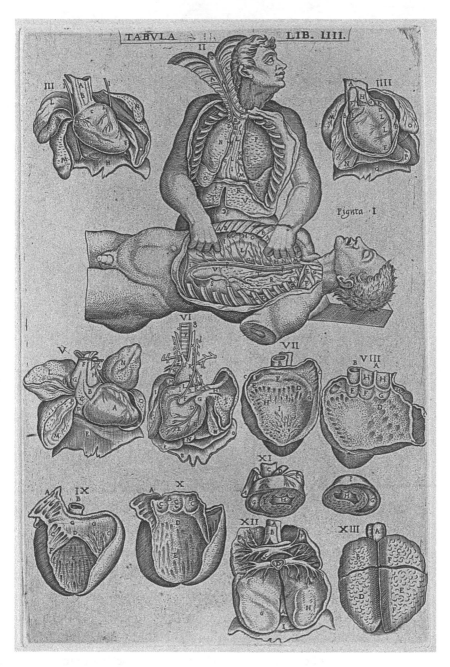

FIGURE 11.2 Two figures with their thoracic cavity exposed, one dissecting the other (figs. I–II), together with illustrations mainly of the heart (figs. III–XI) and two of the lungs (figs. XII–XIII). Engraving, 1568.

Source: Wellcome Collection. Public Domain Mark.

Michael Servetus (1511–1553), a Spanish theologian, physician, and anatomist working in France, "discovered" that the right heart pumped blood to the lungs. This, in fact, was previously described by Arab physician Al-Nafis in the thirteenth century, and Servetus may have read Al-Nafis's work. Servetus wrote this in his book, *Christianismi Restitutio* (The Restoration of Christianity):

> The vital spirit is generated by mixing of the inspired air with blood, which goes from the right to the left ventricle. This blood transfer does not occur through the ventricular septum, as usually believed [Galen's theory that held true for over a thousand years], but through a long conduit crossing the lungs. The blood is refined and brightened by the lungs, goes from the pulmonary artery to the pulmonary veins, is mixed with the inspired air and eliminates residual fumes. Eventually the whole mixture is sucked by the left ventricle during diastole.

Servetus noted that the blood entering the lungs was a different color than that leaving it. We now know this is due to the different oxygen levels in arterial versus venous blood.

Unfortunately, Servetus's life was cut short before he might have made more discoveries about the heart. He wrote extensively on religion, and in 1553 he got into a correspondence fight with John Calvin. Calvin denounced him, and Servetus was deemed a heretic and imprisoned in France. After escaping prison, having been sentenced to burn, he inexplicably went to Geneva to continue the argument with Calvin, where Calvin's supporters condemned him to burn at the stake on a pyre of his own books.

By the middle of the sixteenth century, physicians and scientists were questioning the Galenic heart. They were beginning to understand the workings of the heart. Painters and poets continued to use the heart as the symbol of lovers, love of God, and of courage and faithfulness. Meanwhile, physicians and scientists began to understand what we now accept the heart to be—a blood pump in the center of a circulatory system.

Chapter Twelve

HITHER AND THITHER

It must then be concluded that the blood in the animal body moves around in a circle continuously, and that the action or function of the heart is to accomplish this by pumping.
William Harvey, 1628

WILLIAM HARVEY, son of a farmer, was the personal physician to two English kings. As a medical student, Harvey considered Aristotle his master. Harvey went to Padua, the center of medical science, to study under Hieronymus Fabricius (a student of Vesalius and the discoverer of valves in veins). Based on his experiments, Harvey was the first to mechanistically describe circulation; how the heart "pumps" blood around the body. For his circulation theory, Harvey had an advantage that Galen and others before him did not have: the invention of mechanical pumps. By Harvey's time, hydraulic water pumps for mining and fire-extinguishing were in common use. The metaphor was there for him to comprehend.[1]

Harvey wrote the *Anatomical Study of the Motion of the Heart and of the Blood in Animals* in 1628, but he had discovered "circulation" in 1615. He waited thirteen years before publishing his findings—not in England but in Frankfurt,

Germany—out of fear for his safety. Challenging the Catholic Church's accepted Galenic dogma was considered sacrilegious.

To disprove accepted Galenic theories, Harvey experimented. In one experiment, he tied off a section of artery with two strings and cut it open. He found there was only blood inside, not air or spirits, as Galen had taught. In a second study, Harvey proved that when the pulmonary artery is tied off and the right ventricle filled with water no fluid crosses the heart's septum into the left ventricle through invisible pores. Harvey wrote:

> It has been shown by reason and experiment that by the beat of the ventricles blood flows through the lungs and it is pumped to the whole body. There it passes through pores in the flesh into the veins through which it returns from the periphery . . . finally coming to the vena cava and right auricle. . . . It must then be concluded that the blood in the animal body moves around in a circle continuously, and that the action or function of the heart is to accomplish this by pumping. This is the only reason for the motion and beat of the heart.[2]

Alas, the heart was merely a pump.

But Harvey publicly stated that the heart was the seat of emotions and did not challenge its metaphysical role (possibly out of fear for his life). He did believe that the heart, near the physical center of the body, by virtue of this circulation, distributed warmth to the rest of the body. This is partially correct, but the hypothalamus in the brain is actually the body's temperature regulator. Harvey wrote, "the temperament of the brain is cold and humid and soft; cold that it may temper spirits from the heart; lest the heart be inflamed and swiftly lose its power because in the insane the brain has become hot."[3]

Galen believed that after food was ingested it was transported to the liver, where it was converted into blood. This theory is what most of Western civilization accepted from the time of Galen for the ensuing 1,500 years. Unknown to Harvey and the world, da Vinci had already calculated the amount of blood that moved with each pulse. And if you multiplied that times the number of heartbeats in a day, it was thousands of liters (its actually 7,600 liters per day). One would need to eat a lot of food to produce that much blood. Harvey came to the same conclusion. He determined that the amount of blood moving through the heart in half an hour was greater than the total amount of blood in the body; blood therefore had to be recirculated.

Harvey performed dissections of dogs and humans (executed criminals) to packed amphitheaters to prove his theories. He would lecture in Latin, accompanied by lutes. He would cut open the pulmonary artery of a dog after the heart had been exposed, showering the audience with blood as the right ventricle contracted—what fun! He demonstrated that blood gets pumped by the heart out to the body through the arteries. It then returns to the heart through the veins, where it is pumped into the lungs to extract some vital force. We now know this is oxygen. Blood is then pumped back out to the body by the left ventricle. Blood is recycled, and that could account for the amount of blood the heart needed to pump daily.

Harvey performed tests to show that blood moved away from the heart through arteries and toward the heart through veins. He applied a tourniquet to a human arm just tight enough to block the flow of blood through the veins, without affecting the muscular arteries. When he did this, the section of the arm below the tourniquet swelled, as would be expected if the blood entered the arm but could not leave it. After loosening the tourniquet, he retightened the tourniquet further, blocking flow in both arteries and veins. Blood did not build up in the veins, and the arm did not swell. Furthermore, blood built up in the arteries above the tight tourniquet. Harvey surmised that blood moved "thither by the arteries, hither by the veins."

How the blood moved from the arteries to the veins was unclear. Harvey theorized that there were invisible pores, too small to see, connecting the two vascular systems. We now know these "pores" are capillaries.

■ ■ ■

Viscount Hugh Montgomery fell from a stumbling horse at age ten, smashing his chest against a jutting rock. Ribs fractured, fever ensued, an abscess formed and then burst, leaving a gaping hole in his left chest. The wound healed but remained open. It was intermittently covered by a steel plate as he aged. As unlikely as it seems, the viscount appears to have lived a perfectly healthy life despite the permanent wound. After traveling Europe to sold out crowds in 1659, he returned to London famous at age eighteen. Harvey informed King Charles I of this unusual case. He was instructed to seek out the Viscount and bring him before the king. Upon meeting the young man, Harvey examined the hole in his chest:

> Being now amazed at the at the novelty of the thing, I searched it again and again, and having diligently inquired into all, it was evident that the old and vast ulcer (for want of a skillful physician!) was miraculously healed and skinned over with a membrane on the inside, and guarded with flesh all about the margin of it.

He further described his examination:

> But that fleshy substance by its pulse and the difference of rhythm there of, or time which it kept (and laying one hand upon the wrist, and the other upon the heart), and also by comparing and considering his respirations, I concluded it to be no part of the lungs, but the cone or substance of the heart. I took notice of the motion of the heart; namely, that at the diastole it was drawn in and retracted, and in the systole it came forth and was thrust out; and that the systole was made in the heart when it was sensible at the wrist. Thus, strange as it may seem, I have handled the heart and ventricles, in their own pulsations, in a young and sprightly nobleman, without offence to him.

> I carried the young man himself to the King, that his majesty might with his own eyes behold this wonderful case: that, in a man alive and well, he might, without detriment to the individual, observe the movement of the heart, and, with his

proper hand even touch the ventricles as they contracted. And his most excellent majesty, as well as myself, acknowledged that the heart was without the sense of touch; for the youth never knew when we touched his heart, except by the sight or the sensation he had through the external integument.

After examining the young man's heart, the king said, "Sir, I wish I could perceive the thoughts of some of my nobilities hearts as I have seen your heart." To which young Montgomery replied, "I assure your majesty, before God here present and this company, it shall never entertain any thought against your concerns, but be always full of dutiful affection and steadfast resolution to serve your majesty."[4]

■ ■ ■

By the end of seventeenth century, anatomical knowledge of the heart was surprisingly accurate, and Harvey's theories of a double circuit, made up of the pulmonary and systemic circulations, became widely accepted. It was during the Renaissance that science changed forever our view of the heart. The heart was now thought of as nothing more than a mechanical pump devoid of spiritual significance.

The French philosopher, mathematician, and part-time physiologist René Descartes (1596–1650) was one of the first scholars to accept Harvey's circulation theory.[5] However, Descartes argued that Harvey was incorrect in describing the heart as nothing more than a passive pump. He believed the heart was more akin to a machine-like furnace (think combustion engine).

Descartes located the seat of the soul in the pineal gland, in the middle of the brain: "The parts of the blood which penetrate as far as the brain serve not only to nourish and sustain its substance, but also and primarily to produce in it a certain very fine wind, or rather a very lively and pure flame, which is called the animal spirits."[6] Descartes's views on the heart in *De Homine Figuris* were published posthumously in 1662 (figure 12.1). He had feared that publishing his writings, completed in 1632, would

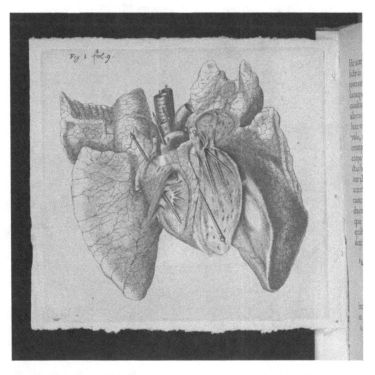

FIGURE 12.1 *De Homine Figuris* (1662), figure 1, heart and lungs with flaps raised, by Rene Descartes.

Source: Reproduced by kind permission of the Syndics of Cambridge University Library.

subject him to an inquisition such as Galileo Galilei had faced in 1633 for his book *Dialogue Concerning the Two Chief World Systems*. The Catholic Church declared that the concept that the Earth rotated around the Sun, and was not the center of the universe, was heresy. Although threatened with torture, Galileo got away with house arrest for the rest of his life.

The heart was no longer considered the seat of the soul, nor was it the place where God communicated with a person. The heart was merely an organ that only responded to the emotions and feelings we have (think of your heart racing when seeing your new love or when a lion is about to pounce on you). Going forward, the heart would only metaphorically be the source of love, courage, and desire—but this metaphor would remain powerful.

PART 3

heART

Chapter Thirteen

THE HEART IN ART

I put my heart and my soul into my work, and have lost my mind in the process.

Vincent van Gogh

If I create from the heart, nearly everything works; if from the head, almost nothing.

Marc Chagall

THE FIRST known artistic illustration of a heart outside anatomical literature in Europe appeared during the Middle Ages around 1255, in *Roman de la poire* (Romance of the Pear). In this illustrated manuscript by a poet named Thibaut, a scene shows a lover kneeling before a damsel offering her his heart; she is looking somewhat taken aback (figure 13.1). Shaped like a pine cone, the heart tip points upward, in accordance with accepted anatomical descriptions of the human heart by Galen and Aristotle. This may be the earliest artistic use of the heart to metaphorically signify romantic love.[1]

In 1305 Giotto di Bondone painted personifications of the vices and virtues on a mural in the Arena Chapel in Padua, Italy. In the upper right corner of the mural, one virtue, Caritas (Divine Love), offers her heart to God above her (figure 13.2). Thus began the artistic depiction of surrendering one's heart to God as a religious symbol of love.

FIGURE 13.1 *Roman de la poire* heart metaphor.

Source: Atelier du Maitre de Bari / Wikimedia Commons / Public Domain.

FIGURE 13.2 Caritas (Charity) of the Seven Virtues in the Arena Chapel in Padua, Italy.

Source: Giotto di Bondone / Wikimedia Commons / Public Domain.

Francesco da Barberino produced an encyclopedic work titled *Documenti d'amore* (The Love Documents) in 1315. Included was an illustrated poem in which Conscientia holds her heart in her hand, which she has torn from her chest to demonstrate that she has a pure conscience—a pure heart. In an illustration in *Tractatus de Amore*, he shows Cupid shooting arrows atop a horse wearing a wreath of hearts (figure 13.3). Thus, the heart could represent purity and virtue and, at the same time, eros and romance. Barberino's epic poem went viral at the time. Within a few years, other artists began illustrating their romantic works with the more decorative, and less anatomical, scalloped heart.

In *The Romance of Alexander*, illustrated by Jehan de Grise in Flanders in 1344, a woman raises a heart received from Alexander the Great, who touches his breast to indicate the place from which it came. By the mid-fourteenth century, heart imagery began to appear throughout Europe, symbolizing seemingly contradictory themes of erotic or romantic love and the unadulterated love of God.

Around this time artists began to depict the heart held by the tip, with the base pointing upward, which was more anatomically correct. The earliest example of this depiction is from the mid-1300s on a small oak chest known as a minnekästchen. The German chest was meant to be a present to one's love for holding jewelry or personal belongings. An illustration on the chest shows a young man presenting his heart to Frau Minne, the personification of courtly love in Middle High German literature. Frau Minne was often referred to as the "Goddess" of romantic love or passion (as opposed to the figure of Jesus who

FIGURE 13.3 Cupid in The Triumph of Love, illustration from *Tractatus de Amore*. The god of Love is shown striking down men and women of various social positions.

Source: Francesco da Barberino / Wikimedia Commons / Public Domain.

represented compassion). A painting from the 1400s discovered in a Zurich guild house in 2009 shows Frau Minne presiding over suffering men in love who are having their upright hearts metaphorically torn from their breasts.[2]

The *Offering of the Heart*, a tapestry by an unknown Flemish weaver from about 1410 hangs in the Louvre in Paris. It is a beautiful example of the chivalric ideal of romantic love. A knight holds his heart, the symbol of his love, between his thumb and forefinger (figure 13.4). The heart looks like what we now recognize as the iconic heart symbol of today.

Frau Minne returned in a print from 1485 by Master Caspar von Regensburg titled *Venus and Her Lover*. Here she is surrounded by no less than nineteen hearts. Shaped like the heart symbol we know today, she looks down at the helpless lover while torturing the hearts in myriad ways (figure 13.5). Hearts are pierced by arrow, knife, and lance. More hearts are clapped in a trap, burned at the stake, sawn in two, and other forms of violent abuse as the lover beseeches her to release him from his pain.

Around 1500, Pierre Sala, a courtier of Louis XII of France, created a tiny book titled *Emblemes et Devises d'amour* (Love Emblems and Mottos) that was meant to be held in the palm of the hand. It contained twelve love poems and illustrations. It was meant for the love of his life, Marguerite Bullioud, the wife of another (they eventually wed when her first husband died). One of the illustrations, "Miniature of Two Women Trying to Catch Flying Hearts in a Net," depicts two women attempting to capture "soaring love," symbolized by hearts with wings (figure 13.6).

FIGURE 13.4 *The Offering of the Heart,* tapestry by unknown Flemish weaver.

Source: Louvre Museum / Wikimedia Commons / Public Domain.

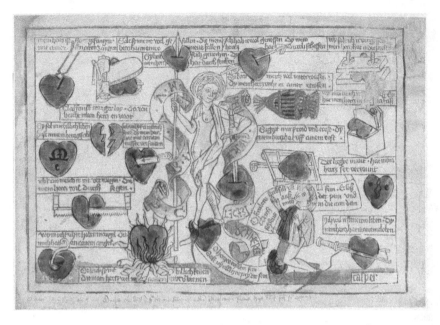

FIGURE 13.5 *Venus and the Lover* [Frau Venus und der Verliebte], c. 1485, Master Caspar.

Source: Bildagentur / Kupferstichkabinett, Staatliche Museen, Berlin, Germany / Jörg P. Anders / Art Resource, New York.

The heart image proliferated throughout Europe in the latter part of the Middle Ages and into the Renaissance. It appeared in the paintings of the time as well as on coats of arms, shields, sword handles, jewelry boxes, and burial stones. Heart-shaped books also became popular, representing the idea of the heart as the seat of memory. Medieval heart-shaped books contained music, such as love ballads, or religious devotions. The unopened book resembled an almond.

FIGURE 13.6 "Miniature of Two Women Trying to Catch Flying Hearts in a Net," from Petit Livre d'Amour, by Pierre Sala.

Source: The British Library [Stowe 955, f. 13].

FIGURE 13.7 *Chansonniere de Jean de Montchenu* (c. 1470) is a book of songs about courtly love (thirty in French and fourteen in Italian).
Source: Courtesy of Bibliothèque Nationale de France.

But upon opening, the book "blossomed" into a heart. Others were shaped like a scalloped heart when closed. When opened they represented two hearts joined by love (figure 13.7).

■ ■ ■

Commercially manufactured playing cards (originally from China, via Egypt, to Europe) began to appear shortly after the invention of the printing press around 1480. The four suits represented the medieval feudal estates. Spades, the

first suit, represented the swords of the gentry. Hearts symbolized the clergy, the "pure-at-heart" (on earlier hand-painted
decks this had been a cup for the Holy Grail). Diamonds
represented merchants, and clubs exemplified agriculture or
the peasantry.

■ ■ ■

Martin Luther was a sixteenth-century monk and theologian whose beliefs helped birth the Reformation, ultimately
making Protestantism the third major force in Christendom
behind Roman Catholicism and Eastern Orthodoxy. Luther
commissioned his own seal in 1530, paid for by his patron John
Frederick of Saxony. The Luther Rose, a white rose surrounding a heart containing a black cross—the heart of Christ the
Crucified—became the symbol for Lutheranism (figure 13.8).
Luther said the cross in the heart was black because it brought
pain. The heart was red because it gave life. The heart rested
on a white rose to show that faith caused joy, consolation, and
peace. Luther stated, "for with the heart man believeth unto
righteousness."

■ ■ ■

In early illustrations of Mesoamerican heart sacrifices, hearts
offered up to the Sun god were portrayed with their tips pointing
up, as on a stele—an upright stone slab monument—from the
Nahuat-speaking Pipil people between 700 and 1200. In later
Mesoamerican depictions, the hearts' tips pointed downward,

FIGURE 13.8 The Luther Rose. I, Daniel Csorfoly (from Budapest, Hungary).

Source: Wikimedia Commons / Public Domain.

with their scallops up, as seen on a 1500 Olmec statue of a man holding a heart (figure 13.9). Is it a coincidence that this artistic change in orientation of the heart coincided with a similar change in Europe?

FIGURE 13.9 Two-part container of man holding human heart. Olmec, Las Bocas, Puebla.

Source: *The Olmec World*, p. 327. Photograph by Justin Kerr, The Pre-Columbian Portfolio / Public Domain.

■ ■ ■

Myths of Erzulie Freda, the Haitian Loa (spirit or goddess) of love and women, date back to West Africa from as early as the sixteenth century. Erzulie Freda became the protector of women for Africans forced into slavery and brought to Haiti by the Middle Passage. Possibly the most popular Loa in Haitian Vodou and New Orleans Voodoo, her symbol is a heart, or a heart pierced with a dagger or sword. Erzulie Freda was a popular subject with Haitian artists, often portrayed as a Madonna with a heart on her chest or a sword stabbed into her chest through her heart. Her Veve (religious symbol) of a heart, often with a sword through it, is still used today in Vodou religious rituals.

■ ■ ■

The heart, symbolic and anatomical, continues to be found in the artistic works of all cultures. Contemporary artists have sustained the popularity of the heart in art. Frida Kahlo used anatomical hearts in her *Two Fridas* (1939), to represent the broken and healthy hearts of her two personalities after her divorce from Diego Rivera (figure 13.10).

The heart represents life and passion in Henri Matisse's *Icarus* (1947). Icarus, a character in Greek mythology who flew too close to the sun, which melted his wax wings, is shown as a black figure falling through the blue sky surrounded by stars (figure 13.11). A bright red oval glows within the breast of Icarus

FIGURE 13.10 *The Two Fridas* by Frida Kahlo. Museo Nacional de Arte Moderno, Mexico City.

Source: Schalkwijk / Art Resource, New York. © 2022 Banco de México Diego Rivera Frida Kahlo Museums Trust, Mexico, D.F. / Artists Rights Society (ARS), New York.

who, as Matisse wrote, "with a passionate heart falls out of the starry sky."

Girl with Balloon is a London series of stencil murals by the graffiti artist Banksy that appeared in 2002. The mural depicts a young girl with her hand extended toward a red heart-shaped

FIGURE 13.11 *Icarus* by Henri Matisse.

Source: Image copyright © The Metropolitan Museum of Art. Image source: Art Resource, New York. © Succession H. Matisse / Artists Rights Society (ARS), New York.

FIGURE 13.12 Banksy Girl and Heart Balloon, Dominic Robinson from Bristol, UK.

Source: Wikimedia Commons / Public Domain.

balloon carried away by the wind (figure 13.12). It is unclear whether the girl has lost her grasp of the balloon, which floats away representing lost hope, or whether she is intentionally releasing the heart, giving hope and love to the world.

The heart remains prevalent in modern art, both symbolically and anatomically. Outside of anatomical drawings, can you recall any paintings of the brain?

■ ■ ■

There are a number of theories as to how the heart symbol became the bright red, scalloped, symmetrical ideograph we recognize today. The modern heart shape is a pictogram—an abstract symbol rather than an anatomically correct rendering. In geometric terms, ♥ is a cardioid, which is common in nature. It could represent the heart-shaped fruit of the silphium plant, a now extinct contraceptive used by the Greeks and Romans in the sixth century BCE. It might be the ivy leaf, frequently seen in ancient Greek art and associated with romantic love. It could also be a representation of a woman's breasts, buttocks, or parted vulva. Others have suggested the image of the courting ritual of swans' necks.

There are many theories, but it may be as simple as a crude portrayal of the anatomical heart. Since the Catholic Church forbade medieval artists from performing dissections, it may be that the symbol we know today is based on the ancient descriptions of Aristotle and Galen, who described the heart as a three-chambered organ with a dent in the middle of the base. More ancient depictions of the heart, such as the heart being weighed in judgment by the Egyptian god Osiris (c. 2500 BCE) or the Mesoamerican Olmec man holding the heart (c. 1500 BCE), are more anatomically correct.[3] These ancients viewed actual human hearts during embalming or sacrifice. These more accurate images of the heart were not discovered until the nineteenth and twentieth centuries, so they weren't available to the medieval artists. Yet when we compare Leonardo da Vinci's drawing of the anatomical heart to the heart ideograph we have come to know as the symbol of love, maybe the heart symbol we recognize today is not so far from the truth (figure 13.13).

■ ■ ■

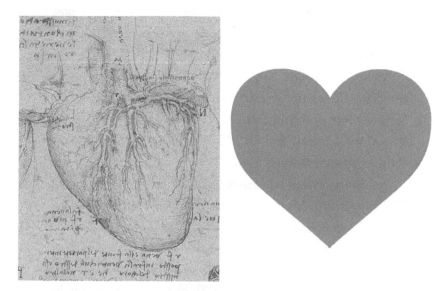

FIGURE 13.13 Da Vinci heart compared to the love heart symbol. Heart drawing and annotations by Leonardo da Vinci.
Source: Collection Windsor Castle, United Kingdom. Royal Collection Trust/ © Her Majesty Queen Elizabeth II 2018.

Graphic designer Milton Glaser created the famous logo I♥NY in 1977 to increase tourism to New York state, and for the first time the heart ideograph became a verb. Now ♥ could be used to love any person, place, or thing: "I ♥ your ♥." Think about that car in front of you with the bumper stickers "Honk if you ♥ Jesus" or "Virginia is for lo♥ers" or that slow driver with the "I ♥ my corgi" one.

Japanese provider NTT DoCoMo released the first emoji in 1995 on its popular Pocket Bell pagers. It was a heart. By 1999 they had developed colored emojis, including five different hearts. Look at your phone to see how many emojis you now have with hearts. There is also a brain emoji. But "I 🖤 NY" and "I 🖤 your " just don't work.

Twitter expanded the meaning and use of ♥ in 2015 by tweeting, "You can say a lot with a heart. Introducing a new way to show how you feel on Twitter." The company wrote: ♥=yes! ♥=congrats! ♥=LOL ♥=adorbs ♥=stay strong ♥=wow ♥=hugs ♥=aww ♥=high five.

The heart symbol remains prevalent in art and social media. The meanings of heart symbols have multiplied; for example, it can now symbolize "health" or "lives" in video games. Despite the disillusionment of the heart as the seat of our soul and our love, the heart shape continues to thrive as a symbol of romantic love, familial love, and love of god. Think of that big guy on the motorcycle that has a heart tattooed on his arm with MOM written across it. Does the girl at the coffee shop not blush when the barista carefully places a latte in front of her upon which he has placed a heart shape in the foam?

Chapter Fourteen

THE HEART IN LITERATURE

In art, the hand can never execute anything higher than the heart can imagine.

Ralph Waldo Emerson

JUST AS the illustrative heart in art appeared during the Middle Ages, so did its symbolic use in the literature of the time.[1] In his autobiographical *Vita Nuova* (1294), Dante Alighieri wrote of his love for Beatrice: "I felt in my heart a loving spirit that was sleeping; and then I saw Love coming from far away so glad, I could just recognize." Dante goes on to dream that Beatrice eats his heart.

In Giovanni Boccaccio's narrative poem *Amorosa Visione* (1342), the theme of the heart as a book, its walls to be written on, shifts from containing words from God to words from one's love:

As I stood there I seemed to see
this gentle woman come toward me
and open me in my breast, then write there
inside my heart, placed to suffer,
her beautiful name with letters of gold
so that is could never leave from there.

In Boccaccio's collection of one hundred tales titled *Decam-eron* (1350), two stories represent the heart, perhaps too vividly, as true love. In the first tale of day four, Tancredi, the Prince of Salerno, is incestuously attracted to his daughter Ghismonda. Out of jealousy, he slays his daughter's lover Guiscardo and sends her his heart in a golden cup. Ghismonda raises her lover's heart to her lips and kisses it. "I know that his soul still resides here in you and is looking at the place where he and I knew happiness." She thanks the servants for her father's price-less gift to her. She adds her tears and poison to the cup, which she drinks. "O beloved heart, all I was to do for you I have now done, and nothing remains but to make my soul join with that one which you have held so dearly." She then crawls into bed clutching her lover's heart to await death.

In the ninth tale of day four, the knight Guillaume de Roussillon murders his wife's lover, his friend who is also a knight. Roussillon cuts out his heart and has it sent to his cook, instructing him to prepare a special dish with this "boar's heart." When the cooked heart arrives at the table, Roussillon has no appetite. He gives the special dish to his lady, who eats up every delicious morsel. Roussillon asks, "Madam, how liked you this dish?" Why yes, she replies, very much so. Roussillon then tells her that he cut the heart out of her lover with his own hands. Instead of gagging and vomiting, the lady declares that she'll never taste another piece of food since she has eaten the most perfect thing in the world. She then steps to the window and jumps to her death.

Shakespeare famously wrote about the symbolic heart in his verse. In his *Sonnet 141* (1609), he writes:

But my five wits nor my five senses can
Dissuade one foolish heart from serving thee.
Who leaves unswayed the likeness of a man
Thy proud heart's slave and vassal wretch to be.

In *Much Ado About Nothing* (1623), after Benedick blurts out that he loves her, Beatrice finally admits, "I love you with so much of my heart that none is left to protest."

And when too proud King Lear (1606) asks his three daughters to profess their love to him to receive parts of his kingdom, unlike her two older deceitful sisters Cordelia (*cor* as in heart) cannot express her love to him: "I cannot heave my heart into my mouth." He does not understand that she cannot express how great her love is for him, and he angrily disinherits her—a Shakespearian tragedy.

The heart as the path to God appears in Sir Arthur Conan Doyle's *The Stark Munro Letters* written in 1895. Dr. Munro states in an argument with the High Church curate of the parish, "I carry my own church about under my own hat, said I. Bricks and mortar won't make a staircase to heaven. I believe with your Master that the human heart is the best temple."

The surest way to rid the world of a vampire can be found in Bram Stoker's *Dracula* written in 1897. Professor Abraham Van Helsing writes to Doctor John Seward that the latter must "Take the papers that are with this, the diaries of Harker and the rest, and read them, and then find this great UnDead, and cut off his head and burn his heart or drive a stake through it, so that the world may rest from him."

The heart has ever since been and continues to be used as a vehicle in the literary arts to represent romantic love, familial love, love of God, of what is good in us. Here are some recent examples in modern literature. In James Joyce's *A Portrait of the Artist as a Young Man* (1916), as a teenager Stephen first experiences desire of the heart: "His heart danced upon her movements like a cork upon a tide." When Sabina in Milan Kundera's *The Unbearable Lightness of Being* (1984) is observing the American senator watching his children run and play, she thinks, "When the heart speaks, the mind finds it indecent to object." In renouncing logic, she is saying that what our feelings in our heart tell us are truer than what our thoughts do.

Wishing that his heart was empty of feelings, trying to disconnect himself from his humanity, the father in Cormac McCarthy's *The Road* (2006) thinks to himself, "If only my heart were stone." In Delia Owens's *Where the Crawdads Sing* (2018), some months after her mother left her, Kya's pain in her heart from loneliness and loss begin to dull: "Until at last, at some unclaimed moment, the heart-pain seeped away like water into sand. Still there, but deep." After Katherine tells him she always loved him, Almásy confesses his tortured heart to Katherine in Michael Ondaatje's *The English Patient* (1992). He must leave her mortally wounded in the cave but promises to come back for her body: "Every night I cut out my heart. But in the morning, it was full again."

Chapter Fifteen

THE HEART IN MUSIC

Music acts like a magic key, to which the most tightly closed heart opens.

Maria Augusta von Trapp

FOLLOWING THE Gregorian chants and liturgical dramas of the Dark Ages, the heart became a regular theme in music during the Renaissance. Songs about love and heartbreak were popular in the fourteenth and fifteenth centuries. Baude Cordier (*cor* as in heart) was the nom de plume of Baude Fresnel (1380–c. 1440), a French composer who wrote rondeaux (refrains) about love. Cordier's rondeau "Belle, Bonne, Sage" was written in a heart shape. In addition to shaping the music as a heart, Cordier substituted the word *heart* with a small drawing of a heart in the lyrics (figure 15.1). The refrain:

Lovely, good, wise, gentle and noble one,
On this day that the year becomes new
I make you a gift of a new song
Within my heart, which presents itself to you.

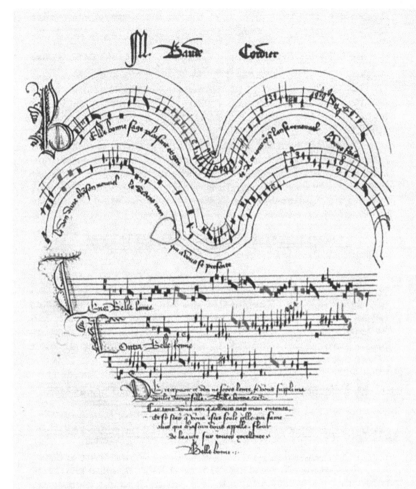

FIGURE 15.1 "Belle, Bonne, Sage" from *The Chantilly Manuscript* by Baude Cordier.

A popular madrigal in Italy from 1601 did not mince words. From Giulio Caccini's "Amarilli mia bella": "To take the arrow out of the lover's heart / Is to heal him of love's wound, /And that can only be done by lovemaking."

"Heart of Oak" was written by the English actor David Garrick in 1759. Garrick is credited with the theatrical blessing "break a leg" as he was once so involved in his performance of Richard III that he did not notice the pain of a fracture he incurred during a play. "Heart of Oak" went on to become the official march of the British Royal Navy. The heart of oak refers to the strongest central wood of the tree—the heartwood. The refrain of the song goes like this:

> Heart of oak are our ships, heart of oak are our men;
> We always are ready, steady boys, steady!
> We'll fight and we'll conquer again and again.

Spirituals, one of the most significant forms of American music, are religious folksongs created by the African slaves in the antebellum American South. The term *spiritual* comes from Ephesians 5:19 in the King James Bible: "Speaking to yourselves in psalms and hymns and spiritual songs, singing and making melody in your heart to the Lord." One of the most famous spirituals from the early 1800s is "Deep Down in My Heart," with these lyrics to live by: "Lord, you know I love everybody, / Deep down in my heart."

The symbolic heart has remained a pervasive metaphor in popular music ever since. Examples of modern heart song titles include: "My Heart Has a Mind of Its Own" by Connie Francis;

"I Left My Heart in San Francisco" by Tony Bennet; "Sgt. Pepper's Lonely Hearts Club Band" by the Beatles; "Un-Break My Heart" by Toni Braxton; "Total Eclipse of the Heart" by Bonnie Tyler; "Don't Go Breaking My Heart" by Elton John and Kiki Dee; "Somebody Already Broke My Heart" by Sade; and "Stop Draggin' My Heart Around" by Stevie Nicks with Tom Petty and the Heartbreakers.

Remember the English rock band The Troggs, and later Jimi Hendrix, belting out how "Wild Thing" made their heart sing and everything groovy. Or how Mick Jagger of the Rolling Stones lamented in "Dear Doctor" how there was a pain where his heart once had been. Lest we forget, there are country greats such as Hank Williams, "Your Cheatin' Heart"; George Strait, "I Cross My Heart"; and Billy Ray Cyrus, "Achy Breaky Heart."

Beginning during the Renaissance and through to Tom Petty and the Heartbreakers and contemporary music, the heart has been used as a symbol in music for romance, love, strength, and heartbreak. *Heart* is the tenth most common word used in pop songs (excluding common words like I, the, and you); the fourth most common word in country music; and the sixth most common word in jazz.[1]

Chapter Sixteen

HEART RITUALS

A hundred hearts would be too few, to carry all my love for you.
Anonymous

THERE IS no tradition more closely associated with the heart than Valentine's Day. The Christian priest Valentine of Rome was martyred by the Romans on February 14, 269 CE, after illegally conducting Christian marriages despite a ban imposed by Claudius Caesar. In the fifth century, to commemorate Saint Valentine, Pope Gelasius declared February fourteenth Saint Valentine's Day. This date came to be associated with the commemoration of romantic love, the day of lovers.

Geoffrey Chaucer's *The Parliament of Fowls* (1381 CE), may be the first reference to the idea that Saint Valentine's Day was a day for lovers:

For this was on Saint Valentine's day,
When every fowl comes there his mate to take,
Of every species that men know, I say,
And then so huge a crowd did they make,

That earth and sea, and tree, and every lake
Was so full, that there was scarcely space
For me to stand, so full was all the place.

Charles, Duke of Orleans, may have sent the first known valentine in 1415 to his wife Bonne of Armagnac (all of sixteen years old). He was being held prisoner in the Tower of London following his capture at the Battle of Agincourt. "I am already sick of love, My very gentle Valentine." Unfortunately, Charles was in prison for twenty-five more years, and Bonne died before he was released. Nevertheless, the practice of sending valentines gradually grew in popularity, with couples exchanging handwritten notes and love tokens.

Into the seventeenth century, the celebration of Valentine's Day in England was limited to those who could afford its rituals. On Valentine's Day, affluent men would draw lots with women's names on them. The man who picked a lady's name was obliged to give her a gift. The earliest English, French, and American valentine cards were little more than a few lines of verse handwritten on a sheet of paper. But over time the writers began embellishing them with drawings and paintings, often including the symbolic heart image. These were folded, sealed with wax, and placed on their intended's doorstep.

A valentine card from 1818 reads: "From him who upon the return of another Valentine's Day, looks forward with pleasure to the time when his hopes may be realized; & at the altar of Hymen [Greek god of love] he shall receive the hand accompanied with the heart of her for whom he feels—not a wild and romantic love, which abates after a short acquaintance but

FIGURE 16.1 Valentine card from 1818.

Source: Courtesy of Hansons Auctioneers.

an affection which time increases rather than diminishes" (figure 16.1).

By the end of the eighteenth century, the first commercial valentine cards appeared in England. They were printed, engraved, or made from woodcuts, sometimes colored by hand. They combined traditional symbols of love—hearts, cupids, flowers—with simple verses such as:

Roses are red
Violets are blue
The first time I saw you
My heart knew

or:

Roses are red
Violets are white
You are my world
My heart's delight

By the 1840s and the Industrial Revolution, mass-produced Valentine's Day cards largely took the place of the handmade variety in England and the United States. With introduction of the penny post in Britain in 1840 and the first postage stamps in the United States in 1847, sending Valentine's Day cards became affordable for ordinary people. The first heart-shaped box of chocolates was introduced by Richard Cadbury in 1861. Hallmark started producing Valentine's Day cards in 1913. In 2019, $20.7 billion was spent on Valentine's Day, with more than one billion Valentine's Day cards and thirty-six million heart-shaped chocolate boxes given.

■ ■ ■

Archaeologists have found evidence of brides' wedding rings in hieroglyphics of the ancient Egyptians 4,800 years ago. Made of sedges, rushes, or reeds, these circles symbolized eternity. Macrobius, in his *Saturnalia* (c. 400 CE)—an encyclopedic celebration of Roman culture—wrote that he learned of the custom of placing the betrothal or wedding ring upon the fourth finger from an Egyptian priest: "Because of this nerve, the newly betrothed places the ring on this finger of his spouse, as though it were a representation of the heart."[1]

One theory holds that the connection between the heart and fourth finger on the left hand was the result of Egyptian physicians observing in patients with pain "in their heart" that the pain would start in their chest and move down their left arm into their fourth and fifth fingers. They concluded the heart and fourth finger must be connected. This ancient observation is an accurate description of classic angina. When the heart muscle is not getting oxygen, pressure or heaviness is felt behind the sternum in the chest, radiating to the left shoulder, and down into the left forearm (along the ulnar nerve), into the pinkie and the lateral half of the ring finger. This is because the heart lies in the left chest, and therefore the pain from the heart radiates to the left side cervical nerve roots. These roots sense pain throughout the left upper extremity of the body. The ancient assumption of a connection between the heart and the ring finger was indeed correct. The wedding ring was worn on the fourth finger of the left hand because it was believed that a nerve or vein ran from there directly to the heart. The Romans called it the *vena amoris* or "vein of love."

Christians first started using a ring in marriage ceremonies during Europe's Dark Ages around 860 CE. During the Christian wedding ceremony, the priest would take the ring and touch in sequence the bride's thumb, index, and middle fingers to represent the Holy Trinity before placing the ring on the fourth finger sealing the marriage.

Although the wearing of wedding rings by brides has been traced back to ancient Egypt, it was only in the latter part of the twentieth century that grooms began doing the same. World War II inspired this change because many soldiers fighting overseas chose to wear wedding rings as a comforting reminder of their wives and families back home.

■ ■ ■

General George Washington of the Continental Army awarded the first Purple Heart to a small number of soldiers in 1782 for "commendable actions" in the American war for independence from Great Britain.[2] The color purple represented courage and bravery. The award and ritual were abandoned for two centuries, until General Douglas MacArthur reinstated it on George Washington's two hundredth birthday in 1932. Today, almost two million Purple Hearts have been given in the name of the president of the United States to any member of the Armed Forces who has been wounded or killed while serving under competent authority in any capacity with one of the U.S. armed services. John F. Kennedy is the only U.S. president who was awarded a Purple Heart; he was seriously injured when the patrol boat he was commanding was sliced in half by a Japanese destroyer in World War II. Even though he was badly hurt, he still directed rescue operations and got his crew to shore. He spent hours swimming in the dark to secure aid and food.

PART 4

HEART 101

Chapter Seventeen

THE PUMP

I'm not a romantic, but even I concede that the heart does not exist solely for the purpose of pumping blood.
Maggie Smith in *Downton Abbey*

Dear heart, please stop getting involved in everything. Your job is to pump blood, that's it.
Anonymous

IMAGINE YOU need a pump to continuously push 1.5 gallons of fluid per minute (about 5.5 liters). Therefore, you require this pump to move about 2,000 gallons (near 7,600 liters) every day. And you want it to do this work without a break for eighty years; that's 730,000 gallons per year, or 58,400,000 gallons over eighty years. It would require 1.5 million barrels to contain all of this fluid. For more perspective, consider that pumping this amount of fluid would be equivalent to running a kitchen faucet turned on all the way for at least fifty years.

To achieve your 1.5 gallon per minute flow, your pump will need to squeeze seventy to eighty times per minute (a little more often than every second), 4,500 times per hour, 108,000 times per day, 39,400,000 times a year, or more than 3 billion times over the eighty years you want the pump to work continuously. But you also never want to have to perform tune ups, maintenance repairs, or inspections. This pump can never fail,

FIGURE 17.1 Heart in hand. Chris from Poznan, Poland.
Source: Wikimedia Commons / Public Domain.

even for a moment; it must be perfectly care free. Oh, and you want this pump to be not much bigger than your clenched fist and weigh less than one pound (figure 17.1).

For efficiency, you need this pump to recirculate 1.5 gallons of fluid, refreshed regularly (about every 120 days), throughout a circuit of tubes every minute. The tubing system you have attached to the pump will be able to cycle the fluid back to the pump. But it's a complicated system (figure 17.2), and if you lay out all the tubing end to end, it would stretch sixty thousand miles (about one hundred thousand kilometers). That's long enough to go around the world more than twice. In one day, this fluid will travel twelve thousand miles, roughly four times the distance across the continental United States.

FIGURE 17.2 Plastinated blood vessels of the human head and torso.
Source: Image copyright © www.vonhagens-plastination.com / Gubener Plastinate GmbH, Germany.

Yet despite the overall length of this tubing system, you need the fluid to recirculate back to the pump in no longer than sixty seconds at its furthest distance from the pump. That would be to the toes and back to the heart, as I am obviously writing about the heart and circulatory system, which includes arteries, veins, and capillaries. The wondrous heart needs to push a red blood cell through the circulatory system in about the time it takes a sprinter to run two hundred meters.

Even at rest the heart muscle works twice as hard as the leg muscles of a competitive sprinter. If you squeeze a tennis ball as tightly as you can in your fist, you are mimicking the effort of one beat of the heart to pump the blood forward. Well-trained athletes can increase their cardiac output (the amount of blood the heart pumps in a minute) seven times, from five to thirty-five liters per minute—that's more than nine gallons per minute! The force of the left ventricle's contraction pushes blood through the sixty thousand miles of blood vessels in the human body with the strength to squirt water five feet into the air out of a garden hose. The heart generates enough energy in a day to drive a car twenty miles.

The experience of holding a heart in one's own hand in a patient's open chest is ineffable. Basically, it feels like strong, thick muscle. You are squeezing it to keep blood circulating (called open heart massage). Suddenly you feel the heart start beating against your palm and fingers. Slowly at first, then quickening with increased force. When I first experienced this miracle, I was in awe. Even that broken heart felt so strong pumping in my hand. And if you remove a heart from the body, as in a heart transplant case, it will continue to beat for up to fifteen minutes, until it has finally run out of oxygen and energy.

One can only imagine how ancient peoples saw this spontaneously beating organ that meant life. They surmised, if the heart was life, it surely must house the soul.

Chapter Eighteen

HEART ANATOMY

Even artichokes have hearts.
Amelie

Everybody has a heart, except some people.
Bette Davis

A BLUE WHALE heart weighs as much as one thousand pounds. It can pump out fifty-eight gallons of blood with every heart-beat, at a rate of eight to ten beats per minute (figure 18.1). An adult human female heart weighs approximately half a pound; a male heart about two-thirds of a pound. A human heart pumps 0.02 gallons (1/3 cup) of blood with each beat. The heart of the world's smallest mammal, the Etruscan shrew, weighs 0.00005 pounds. It beats up to 1,511 times per minute. Interestingly, the Etruscan shrew only lives for one year; the blue whale for 80 to 110 years. The creature with the smallest heart on Earth is the *Alpatus magnimius*. This fruit fly is less than 0.01 inches long. You need a microscope to see its heart. Octopuses and squids have three hearts. Hagfish, sometimes referred to as slime eels, have four hearts. Earthworms have five.

FIGURE 18.1 Blue whale heart.

Source: dpa picture alliance / Alamy Stock Photo.

The earliest known heart and blood vessels were found in a fossil from 520 million years ago. The fossil of an arthropod, *Fuxianhuia protensa* (think prehistoric shrimp) was found in the Chengjiang fossil site in southwest China, dating from the Cambrian era.[1] Its tubular heart was in the back of the animal with paired blood vessels extending through the body segments, ultimately concentrating around the brain, eyes, and antennae, where most of the nutrients and oxygen would be required. This anatomy was so successful half a billion years ago that it can still be found in arthropods—invertebrate animals having an exoskeleton, a segmented body, and paired jointed appendages, including insects, arachnids, myriapods, and crustaceans—the oldest and largest phylum today.

Fish have two-chambered hearts. Reptiles have three-chambered hearts with two atria and one ventricle—the exception being alligators and crocodiles, which have four. Spiders have a straight tube for a heart, and a cockroach heart has thirteen chambers.

Birds and mammals, including most humans, have four-chambered hearts: two atria and two ventricles. A rare human birth defect is a three-chambered heart. *Atrium* is Latin for "entrance hall" or "gathering place." The atria are the upper collecting chambers of the heart, receiving blood from the lungs and the body. Ventricle comes from the Latin *ventriculus*, which means "belly." The ventricles, under the atria, are the muscular pumping chambers of the heart, pushing blood out to the lungs and the rest of the body.

■ ■ ■

A penis with a heart.
Biologist Stephen Jay Gould

No, this is not what you're thinking. The deep sea, hideous-appearing anglerfish—the translucent fish with fanged jaws and a reading lamp hanging off its forehead to seduce dinner—practices a unique mating ritual called "sexual parasitism."[2] The tiny male attaches to the female by fusing his circulatory system to hers.[3] His internal organs, fins, eyes, and most all of his body degenerate until nothing is left but a two-chambered heart supplying blood to a penis. He gets nutrition from her blood, she gets on-demand semen.

■ ■ ■

The human heart is actually two pumps. The right and left sides of the heart have different jobs. The right side of the heart pumps oxygen-depleted blood, returning from the body, to the lungs. The lungs load red blood cells with oxygen, which then flow to the left heart. The left side of the heart pumps that now oxygen-rich blood to the rest of the body. The heart continuously pumps this oxygenated blood to almost all seventy-five trillion cells in your body. Only your corneas receive no blood supply.

The specific circuit that blood travels through the heart is as follows: blood, depleted of oxygen and now carrying carbon dioxide, returns from the entire body into the right heart through the venous system (figure 18.2). This deoxygenated blood first enters the right atrium, coming in from the largest veins in the

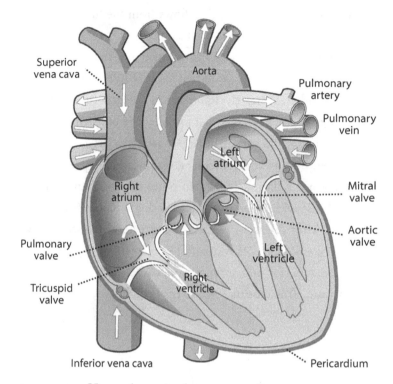

FIGURE 18.2 Human heart circulatory system diagram.
Source: Wapcaplet / Wikimedia Commons / Public Domain.

body, the superior and inferior vena cavae. This blood is then sucked and pushed into the right ventricle through the open tricuspid valve (heart valves open and close to keep blood flowing forward and not backward). The right ventricle pushes the blood through the pulmonic valve into the pulmonary artery and on to the lungs, where blood capillaries sit on 300 million air sacs. At the blood-air interface, hemoglobin—a protein in red blood cells—releases carbon dioxide and takes up oxygen.

The now oxygenated blood then flows from the lungs, through the pulmonary veins, into the left atrium.

This blood is next sucked and pumped into the left ventricle through the mitral valve, so named because it resembles a bishop's miter or hat. When the left ventricle squeezes (or contracts), oxygenated blood is pushed out to the entire body through the aortic valve into the aorta (the largest artery, as large as a garden hose), then into arteries, arterioles, and eventually passing into 600 million capillaries—tubes ten times smaller than a human hair and just wide enough to allow a single blood cell to pass through. After releasing oxygen to organs and tissues, deoxygenated blood makes its way back to the heart through venules, veins, the superior and inferior vena cavae, and back into the right atrium. This repeats 108,000 times daily.

Chapter Nineteen

HEART SOUNDS

EDGAR ALLAN POE'S short story *The Tell-Tale Heart* (1843) ends: "'Villains!' I shrieked, 'dissemble no more! I admit the deed!— tear up the planks! here, here!—It is the beating of his hideous heart!'"

The narrator, who is mad, has murdered and dismembered an old man. As he sits speaking with the police (on a chair over floor boards under which the body parts are hidden), the murderer hears the old man's heart start beating, at first "a low, dull, quick sound—much such a sound as a watch makes when enveloped in cotton." The sound of the beating heart grows louder and louder until he can no longer stand it and confesses to his crime. Does the beating of the heart symbolize the narrator's guilt? Is it actually his own heart that he hears beating?

Poe's metaphoric description of the beating heart, "a sound as a watch makes when enveloped in cotton," is that "thump-thump" or "lub-DUP" many verbalize to describe the heartbeat in English. In Italian it is "tu-tump"; in Polish "bum-bum"; in Norwegian "dunk-dunk"; in Arabic "tum-tum" or "ratama-ratama"; in Nepali "dhuk-dhuk"; in Tamil "lappu-tappu"; in Malay "dup-dap"; and in Indian "dadak" (Hindi) or '"hak-dhak" (Urdu).

Most believe the heart sounds are the pounding of the heart beating. That English lup-DUP sound is actually made by the heart valves snapping shut. The first heart sound, the "lup," is produced by the simultaneous closures of the mitral and tricuspid valves (the valves between the atria and ventricles), which occurs when the ventricles contract and the back pressure forces them to clap shut. The second heart sound, the "DUP," is produced by near simultaneous closures of the aortic and pulmonic valves (the valves through which blood is pumped out of the ventricles to the body and lungs). These two sounds are referred to as S1 and S2.

The stethoscope (which we will learn more about later) can transmit a number of other heart sounds to the ear. These include murmurs, clicks, snaps, knocks, rubs, gallops, and plops. Plops are not good, and fortunately are rare, as they are caused by a tumor plopping back and forth through a heart valve. Murmurs are common and can be innocent (benign) or abnormal (pathological). Heart murmurs are sounds generated by a turbulent flow of blood—think of listening to a babbling brook, rapids in a river, or the distant rumble of thunder. Heart valve stiffening and not opening completely (stenosis) or failing to close correctly and leaking backward (regurgitation) are typical causes of murmurs. These can be mild and of no clinical significance or severe and require surgical replacement or repair; if not corrected, they may result in end of life.

Other abnormal heart sounds can signal valve problems, holes between heart chambers, fluid around the heart, or heart failure. These sounds can result from congenital abnormalities—that is, being born with a heart defect—such as septal defects (holes between the left and right sides of the heart). They can also result from developed or acquired abnormalities. For example, if you're sleeping under a thatched roof in Central or South America, you may wake up to find you have been bitten during the night by a kissing bug that has fallen from the roof. The bug carries a parasite called Trypanosoma cruzi. Infection by this parasite can lead to acute myopericarditis, or injury to the heart muscle and surrounding heart sac. Stethoscope examination can reveal new heart sounds, including gallops, rubs, and murmurs.

Although ancient peoples did not have stethoscopes to hear heart sounds—stethoscopes were not invented until the 1800s—they did listen to the beating of the heart by placing their ear to the chest of patients, animals, partners, and children. When they heard that continuous beating sound that signified life, they knew it came from the heart. They had learned that these sounds came from the heart through hunting and sacrificing animals, embalming bodies, and performing dissections and vivisections. As civilizations developed, heart sounds inspired thinkers, theologians, and philosophers to contemplate existence and what made us "us." They wondered about the location of our emotions and thoughts, our consciousness, and our center of being. Most ancient thinkers concluded that this center was in the heart because the heart beat faster and stronger when we felt love or hate, or had good or evil thoughts. But the heart had competition as the repository of consciousness and the soul. A person lost consciousness when struck in the head, not in the chest. As early as ancient Greece, the cerebrocentrists were arguing that the brain, not the heart, ruled our thoughts and emotions. This heart-brain competition continues to this day.

Chapter Twenty

THE COLOR OF BLOOD

The blood is the life.
Bram Stoker's Dracula

Blood will have blood.
Shakespeare's Macbeth

BLOOD IS the essence of life. Blood is fertility. The Maasai people of southern Kenya drink cows blood at the birth of a baby or the marriage of a daughter (it is also given to drunken elders to alleviate a hangover). For most ancient civilizations, blood carried "life" to all parts of the body.

Blood is also pain and suffering. In Christianity, the blood of Christ represents atonement for humanity. Blood is the food of supernatural beings. Blood is family—remember blood is thicker than water.

Human blood comes in different shades of red, but it is always red because red blood cells contain the protein hemoglobin, which contains iron. Oxygen-rich blood in the arteries is bright red or crimson. Deoxygenated blood in the veins is dark red or maroon. Approximately 8 percent of one's total weight is blood. That would be twenty-five trillion red blood cells with 270 million molecules of hemoglobin in each cell.

Each hemoglobin has room for four oxygen molecules. That's more than one billion oxygen molecules in each red blood cell. An average-sized woman has approximately nine pints of blood and an average-sized man about twelve pints. Losing 20 percent of one's blood volume results in hemorrhagic shock.

Your veins are not blue. They are a dark reddish-brown because of the deoxygenated blood running through them. They appear blue because the skin and underlying fat scatter the red light, allowing only the blue light to travel down to the veins. Because blue is the only color of light that makes it to the veins, it is the only color that is reflected back and, thus, veins appear blue. People who are suffocating turn "blue" because the blood under the skin becomes oxygen-depleted, giving the body a bluish cast because of this light-scattering phenomenon.

So, then, who are "blue bloods"? Blue blood is a literal translation of the Spanish *sangre azul*. This designation was attributed to the oldest and proudest families of Castile in the seventeenth century. The nobles pretentiously claimed never to have intermarried with Moors, Jews, or other races. The expression probably originated because of the noticeable blueness of the veins of these royals of fair (pale) complexion when compared to the common people who had darker skin colors. This term eventually came to refer to all European nobility. It was a metaphor that described the blue appearance of the veins and skin.

Royals coveted the notion that they were "untouched by the sun" and thus were not "darkened." Tanning was a sign of a manual laborer. White, almost translucent skin became a part of the royal image—a sign of beauty. An additional theory holds that leeching of silver from tableware—one only drank wine from a silver chalice—caused the condition argyria (from the Greek word for silver, *argyros*). Argyria can cause the skin to turn a bluish gray.

There is such a thing as blue blood, but you have to be a crustacean, spider, scorpion, horseshoe crab, squid, octopus, or other mollusk to have it. Instead of iron-containing hemoglobin, their respiratory pigment is copper-containing hemocyanin.

Leeches and worms, as well as a New Guinea lizard, have green blood. Invertebrate animals, such as sea squirts, have yellow blood. Other marine invertebrates, such as the unfortunately named penis worm, have violet blood. And finally, the crocodile icefish has blood that is transparent.

For most ancient civilizations, blood carried life to all parts of the body. Blood was almost as important as the heart in many sacrifices to the gods, whether ancient Vikings gathering sacrificed horse blood in a bowl to fling onto altars and participants, the Mayan ritual bloodletting to be smeared on idols of the gods, or the modern day young Maasai male drinking the blood of a bull during his coming-of-age ceremony.

Chapter Twenty-One

THE HEART'S ELECTRICAL SYSTEM

Pain reaches the heart with electrical speed, but truth moves to the heart as slowly as a glacier.
Barbara Kingsolver, *Animal Dreams*

There are two types of cardiologists: plumbers and electricians.
Vincent M. Figueredo, MD

WHEN I talk to patients who have electrical problems with their heart—abnormal heart rhythms or a faltering heart rhythm— I tell them there are two types of cardiologists: plumbers and electricians. I am a heart plumber, and what they need is a heart electrician. I refer them to one of my cardiology partners who specialize in treating arrhythmias, placing pacemakers and defibrillators—the electrophysiologist.

Carlo Matteucci, an Italian physicist, connected a frog's heart to its leg muscle and found that the muscle twitched with each heartbeat. He surmised in 1842 that an electric current accompanied each heartbeat. The heart had its own electrical system![1]

Without an electrical current spreading throughout the heart muscle, there would be no heartbeat and no heart pumping. Electrical impulses from specialized heart cells that act as mini-pacemakers stimulate the rest of the heart muscle cells (myocytes) to contract and to propagate the electrical impulse to neighboring myocytes. Synchronized contraction of myocytes results in the coordinated squeezing of the whole heart muscle to effectively eject blood to the next heart chamber, the lungs, or out to the rest of the body.

The electrical voltages of the heart's pacemaker cells change as charged ions (sodium, calcium, and potassium) flow in and out of the cells' walls, periodically arriving at a voltage threshold that triggers an electrical impulse. This happens about once a second or second and a half in a normal person at rest. The heart beats an average of 78 beats per minute (BPM) in women and 70 BPM in men. At birth, a human baby heart rate averages 130 BPM. An elephant's heart beats at 25 BPM, and a canary's heart beats at 1,000 BPM.

The electrical system in the human heart cycles approximately seventy-five times per minute, 4,500 times per hour, 108,000 times per day, 39,400,000 times in a year, and greater than 3 billion times over eighty years of life.

Amazingly, the heart's electrical system usually functions without fail for our entire lives. But until the mid-twentieth century, if your heart's own pacemaker gave out, that was the end of you. Our lives are now prolonged with mechanical pacemakers no bigger than a matchbox (some now as small as a vitamin pill). The average age of pacemaker recipients is seventy-five years of age. Although the heart's natural pacemaker tends to last seventy-five years and beyond, most mechanical pacemaker batteries only last six to ten years before requiring replacement. As an alternative, scientists are now working on implantable biological pacemaker cells.

Chapter Twenty-Two

WHAT IS AN EKG?

AUGUSTUS DESIRÉ WALLER measured the electrical activity of his dog Jimmy's heart in 1887. During his lectures on the electrical system of the heart, he placed Jimmy's limbs into basins of salt water and then connected the basins to an electrometer (figure 22.1). The electrometer projected an image of the heartbeat onto a photographic plate fixed to a passing model train. The resulting graph showed a wave pattern he called an electrogram—a telegram from the heart. Although crude, these were the first recordings of the heart's electrical system.[1]

The heart beats because heart muscle cells contract rhythmically. They do so because of electrical impulses sent to them by other cells in the heart called pacemaker cells. This electrical current is strong enough to be measurable. The resulting signal can be recorded on running paper (or now on a computer screen), the electrocardiogram or ECG. Most know it as an EKG, from the German *elektrokardiogramm. Elektro* comes

FIGURE 22.1 Jimmy the dog with electrodes attached.
Source: Wellcome Collection, Attribution 4.0 International (CC BY 4.0).

from the Greek for "amber" that, like electricity, was believed to have the power to attract; *kardio* from the Greek for "heart"; and *gramm* from Greek for "drawing" or "writing."

Willem Einthoven first used the term *electrocardiogram*. He published the first electrocardiogram resembling what we now recognize as the modern EKG in 1902. He did not want to use Waller's terminology for the "bumps" on the EKG (A, B, C, D, E), so he instead used P, Q, R, S, T, U, which is what we use today. He devised a way to accurately measure the heart muscle's electrical activity using a device called a string galvanometer. This EKG machine consisted of a series of long, extremely thin, silver-coated glass filaments capable of conducting electricity generated by the heart. To create thin enough glass filaments, Einthoven attached molten glass to an arrow that he fired across the laboratory, stretching the glass into extremely fine wires. He then placed the silver-coated wires into a powerful magnetic field. When the wires were attached to jars of saline into which a patient's arms and left leg were bathed (to enhance conduction), the magnets would bend them to different degrees according to the heart current they carried. This displacement was projected onto a photographic plate, creating a spiked graph. Einthoven's initial EKG machine weighed six hundred pounds, required five operators, and had to be continuously cooled with water to prevent the electromagnets from overheating. Today EKGs weigh a few pounds in the doctor's office and practically nothing on your smart watch.

Einthoven started transmitting EKGs in 1905 from the hospital to his laboratory 1.5 kilometers away via telephone cables—the first telemedicine! Einthoven went on to receive the Nobel Prize for Physiology and Medicine in 1924 for the invention of the electrocardiograph. Led by Horace Darwin (Charles's youngest son), the Cambridge Scientific Company was one of several manufacturers producing commercial versions of EKG machines by the 1930s.

The electrical system of the heart is set up to maximize its pumping action. The main pacemaker in the heart is the sinoatrial node, positioned high in the right atrium (figure 22.2). It rhythmically sends out electrical impulses one to one and

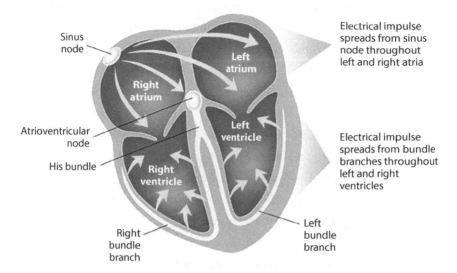

FIGURE 22.2 Electrical conduction system of the heart.
Source: Copyright © 2022 UpToDate, Inc.

a half times per second. The electrical impulses travel through the right and left atria first, which causes heart muscle cells to contract. That causes the atria to squeeze blood forward into the ventricles. The electrical signal from the atria then collects in another pacemaker, the atrioventricular node, positioned between the atria and ventricles. The atrioventricular node delays the electrical impulse long enough for the ventricles to fill with blood. Then it sends the electrical impulse on to the heart muscle cells of the ventricles so they can squeeze the blood forward, out the aorta and pulmonary artery.

■ ■ ■

If the heart had an electrical system to maintain its function that could be recorded as an EKG, could external electrical stimulation revive a nonbeating heart? The Danish physicist Nickolev Abildgaard placed electrodes on the sides of a chicken's head in 1775. He applied an electric charge, which caused the chicken to fall dead. Repeated discharges over the chicken's body had no effect until the electrodes were placed across the chest. Upon an electric discharge across its breast, the chicken got to its feet and staggered away. The first heart defibrillation!

Alexander von Humbolt, a Prussian polymath, attempted to revive a dead finch in 1792 by inserting silver electrodes into its beak and rectum, then applying a current. The bird opened its eyes and flapped its wings before dying again a few minutes later. He then tried the exact same experiment on himself. That didn't go so well.

The German physician Hugo von Ziemssen encountered a forty-six-year-old woman in 1882 who had had a chest tumor removed, exposing her heart through a thin layer of skin. He found that he could change her heart rate by applying electrical impulses to the heart surface. She did not feel the shocks but did feel her heart beating faster.

Albert Hyman, a cardiologist, recognizing that the heartbeat was an electrical phenomenon, developed a technique to shock hearts that had stopped by stabbing an electrically charged gold-plated needle into the patient's chest and into the right atrium of the heart. Jump starting the heart worked! Hyman's device, developed in 1932, was powered by a hand-cranked motor that he called an "artificial pacemaker," a term still used today.

Wilson Greatbatch, an electrical engineer, developed a way to pace a heart whose electrical system was no longer working, resulting in an implantable pacemaker by 1960. Not long after that, in 1980 Michel Mirowski developed and implanted the first automatic defibrillator, which could shock a person at risk out of a life-threatening arrhythmia (e.g., ventricular tachycardia) no matter where the person was.

One of my patients, a deer hunter, was several miles into the woods when the shock of his defibrillator knocked him to the ground. He was able to get up and walk back to his truck, already several miles into the woods, and drive to a local hospital. The implanted defibrillator saved his life. He continues to walk the woods miles from any hospital or anyone.

Chapter Twenty-Three

WHAT IS BLOOD PRESSURE?

My doctors told me this morning my blood pressure is down so
low that I can start reading the newspapers.
Ronald Reagan

SAMUEL SIEGFRIED KARL RITTER VON BASCH, an Austrian-Jewish
physician, best known as the personal physician of Emperor
Maximillian of Mexico, invented the sphygmomanometer in
1891—yes, that is what your blood pressure monitor is called.[1]
Sphygmos comes from Greek for "pulse," *manos* from Greek for
"thin" or "rare," and *metron* from Greek for "measuring."

Blood pressure is the pressure that blood exerts on the inside
of arteries throughout the body. When the heart beats, it creates
pressure to push blood through the arterial tree to get oxygenated
blood to every cell in the body (except the cornea). When the heart
is relaxed between beats, the pressure produced by the blood in
the arteries is the diastolic (from the Greek "to separate") pressure.
When the heart is pumping, the pressure exerted is the systolic
(from the Greek "to pull together") pressure. Blood will squirt
thirty feet up into the air from the pressure created by a heartbeat.

Normal blood pressure in humans is less than 120 mmHg systolic over less than 80 mmHg diastolic; that is 120/80. Your dog's blood pressure should be around 130/75; your cat 130/80. Mice come in at 120/70; a horse at 110/70; and an elephant at 180/120. The mammal with the highest normal blood pressure— due to a distance of six feet between their heart and brain—is the giraffe at 280/180.

Nearly half of adults in the United States (and one in four worldwide) have high blood pressure or "hypertension."[2] High blood pressure—often referred to as the silent killer because the first sign can be a life-ending stroke or heart attack—is when these pressures (systolic and diastolic) are consistently elevated above a healthy pressure of 120/80. Causes of high blood pressure include genetics (thanks to mom or dad), age, obesity, smoking, alcohol, salt, physical inactivity, diabetes, and kidney disease.

Over time the resulting force and friction of elevated pressures on the inner lining of the arteries causes damage. Cholesterol creeps into the damaged artery wall, creating atherosclerotic plaques. The long-term results of untreated high blood pressure are heart attack or heart failure, stroke, kidney failure, peripheral artery disease, and sexual dysfunction.

The results of high blood pressure were recognized in ancient times. Ancient Egyptian, Chinese, and Indian medical texts described patients with a "hard and bounding pulse." These patients did not survive long. Recommended treatments included bloodletting or bleeding by leeches to lower the tension.

Franklin Delano Roosevelt (FDR) is a classic case study in everything that can go wrong with untreated hypertension.[3] At the beginning of his presidency in 1933, FDR already had mild hypertension by today's guideline recommendations, around 140/90 mmHg. By 1944, his blood pressures were over 200/120 mmHg, and he was manifesting heart failure symptoms. At the Yalta conference, with blood pressures of 250/150 mmHg, FDR was audibly wheezing and unable to complete sentences in radio addresses, suggesting severe heart failure. Some historians believe that Stalin took advantage of a debilitated president, determining the fate of Eastern Europe. In April 1945, sitting for a portrait he suddenly complained of the worst headache of his life and fell unconscious. His last measured blood pressure was 350/195 mmHg, and he died of an intracerebral hemorrhage—a ruptured blood vessel causing bleeding inside the brain. Unfortunately for FDR, the first effective and tolerable antihypertensive medications did not become available until the 1950s.

Why is high blood pressure still referred to as "essential" hypertension? Essential for what other than heart attacks and strokes? At the turn of the last century, doctors believed it was essential that the blood pressure be elevated to perfuse critical organs, such as the brain and kidneys, once the arteries started hardening due to atherosclerosis. As Sir William Osler, a giant in medicine, stated in 1912, "the extra pressure is a necessity—as purely a mechanical affair as in any great irrigation system with old encrusted mains and weedy channels."[4] We now know this not to be the case, but the terminology has yet to change.

Treating high blood pressure has been a career interest of mine. I trained to be a "hypertension specialist" and to understand the mechanisms underlying high blood pressure and how to better treat it. When starting medical therapy, my patients universally state, "I don't feel bad." It's my mission to help them understand hypertension's role in heart attacks, heart failure, and strokes. They are often surprised by how much better they actually feel when we "fix" their blood pressure. And I know that I contributed to extending their healthy life.

Lowering blood pressure reduces the risk of heart attacks 25 percent, strokes 35 percent, and heart failure 50 percent.[5] So, as the American Heart Association stresses—and yes, stress can add to high blood pressure—know your numbers and make changes that matter!

Chapter Twenty-Four

WHAT IS HEART FAILURE?

*If the human heart sometimes finds moments of pause as it
ascends the slopes of affection, it rarely halts on the way down.*
Honoré de Balzac, *Pere Goriot*, 1835

*The heart of a mother is a deep abyss at the bottom of which you
will always find forgiveness.*
Honoré de Balzac

HONORÉ DE BALZAC, the nineteenth-century French novelist and
playwright who wrote *La Comédie Humaine*, about French life
in the years after the fall of Napoléon Bonaparte, had "congestive heart failure."[1] Due to a "failing" heart, Balzac's body accumulated fluid, and his legs became grossly edematous (swollen
with fluid). His friend Victor Hugo (*Les Misérables* and *Hunch-
back of Notre-Dame*) wrote that Balzac's legs resembled salty
lard. His physicians tried to drain his fluid-filled legs by sticking
metal tubes through the stretched skin, which was likely already
infected (cellulitis). Gangrene quickly ensued, and Balzac died
shortly thereafter at fifty-one years of age.

Myriad causes (etiologies) can damage organs, including
infections, chemical insults, trauma, or lack of blood supply. The
final common result is organ failure: liver failure, kidney failure,
and when the wheels come off, multi-organ failure.

Heart failure occurs when the pump stops working well for whatever reason. Most commonly this is due to athero-sclerotic coronary artery disease and resulting heart attacks, but other causes are alcohol abuse, viral infections, heart valve problems, and some chemotherapies (the list goes on). A poor pump causes reduced blood flow to the cells of the body due to decreased blood pressure. To keep getting oxygenated blood to cells and tissues, the body has to get the blood pressure up. Hormones are released that increase the heart rate and force the kidneys to retain fluid to increase blood volume (and there-fore blood pressure). Although this works for a time, the body becomes "congested" as fluid accumulates and leaks into tissues; hence the term *congestive heart failure.*

Eventually, the body fills with fluid, especially in the legs (the result of gravity), as in Balzac's case, as well as in the abdomen and lungs. Medications have been developed to prevent this end result for a time, but without healing the heart (sometimes possible, sometimes too late) congestion progresses. In the twenty-first century, new treatments are now available to reduce heart failure symptoms and death. Ventricular assist devices and heart transplants are now routine. Research is looking at injecting healthy cells into the failing heart to rebuild muscle. Physician-scientists are studying xenotransplantation—yes, placing another animal's heart in a human.

Possibly the oldest discovered case of heart failure, more than 3,500 years ago, was in an Egyptian dignitary named Nebiri during the reign of Pharaoh Thutmose III (1424 BCE).[2] Originally discovered in a plundered tomb in 1904 in the Valley of Queens, Luxor, Nebiri's head and various organs that had been placed in canopic jars indicated that he was forty-five to sixty years of age when he died. On examination of the lungs, fluid accumulation was observed in the air spaces of the lungs, suggesting pulmonary edema and heart failure.

As early as the fourteenth century, "dropsy" (from the old French *hydropsie*, from the Greek *hydrops* or *hydro*, which means water), which we now know as congestive heart failure, meant the end of life was near.[3] The body swelled and sufferers eventually drowned from fluid accumulating in their lungs or infections in their swollen legs. The failing heart as a cause was not yet understood. First, physician-scientists needed to understand the heart's role in circulating blood and moving fluid to all parts of the body. The ancient Egyptians, Chinese, and Indians first suggested it. This concept of circulation was the first step in understanding the heart's role in managing fluids in the body. It wasn't until the seventeenth century that William Harvey, who was willing to risk his life and go against church doctrine, stated the case for the pumping heart and its role in circulation.

Chapter Twenty-Five

WHAT IS "HAVING A CORONARY"?

You know I used to be a heartthrob and now I'm a coronary.
Davy Jones of the Monkees

DESCRIBED AS EARLY AS the sixth century BCE in the ancient Ayurvedic text *Sushruta Samhita* as *hritshoola*, meaning "heart thorn"—the Greeks called it "lightning in the chest," and William Heberden first called it "angina pectoris" in 1768— today is often referred to as "having a coronary." The word *coronary* comes from the Latin *coronarius* or "of a crown." The coronary arteries encircle the heart as a crown would on the head of a queen or king.

The coronary arteries, which supply the heart muscle with oxygenated blood, were present in our ancestral lungfish. They supply the hearts of all species with oxygen and nutrients. In mammals and birds, the sizes of the coronary arteries are increased, as are the webs of their branches (arterioles and capillaries), which feed every heart muscle cell. But as the lungfish had two coronary arteries 100 million years ago, so do we still.

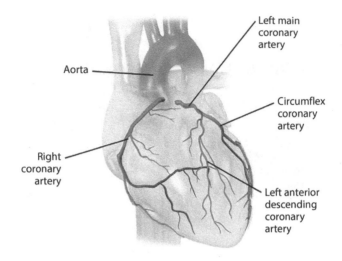

FIGURE 25.1 Coronary arteries.

Source: BruceBlaus / Wikimedia Commons / Public Domain.

It was easier for evolution to enlarge these arteries and their branching instead of making more of them.

The left and the right coronary arteries come off the aorta just above the aortic valve. These two arteries branch into smaller arteries to bring oxygen and nutrients to the entire heart muscle (figure 25.1). If any of them become blocked by arteriosclerosis (cholesterol plaque) and blood clot, death of the region of heart muscle that artery feeds ensues. This is why an acute myocardial infarction (MI) or "heart attack" is sometimes referred to as having a coronary. The afflicted suffer chest pain, shortness of breath, and sometimes sudden death due to dangerous rhythms, including ventricular fibrillation or "vfib." If the heart attack victim survives but the coronary artery was not opened up soon enough, the affected region of heart muscle is irreversibly replaced by scar tissue. Today more than one in three adults in the world, both women and men, will die of cardiovascular disease, mostly heart attacks. Why are so many people having coronaries?

Surprisingly, significant cholesterol plaque buildup in the coronary arteries was found on autopsies of young U.S. soldiers killed in action in Korea in 1953.[1] Their average age was twenty-two. This was later corroborated by casualties of the Vietnam war (mean age twenty-six)[2] and in young victims of violent deaths (median age twenty).[3] But these are kids! It turns out fatty streaks can start forming on the inner walls of the coronary arteries as early as our teens. And with the way we eat now, these fatty streaks become calcified cholesterol plaques that continue to grow in the arterial walls. This was called "hardening of the arteries" by the physician James Herrick in 1912.[4]

These cholesterol plaques become volcanos on the inner lining of the arterial wall waiting to erupt. If the plaque's hard calcium cap cracks and the fatty cholesterol core leaks into the blood, the body's coagulation system—meant to protect the body from uncontrolled bleeding after an injury—mistakenly activates. Platelets from the passing blood attack to quickly form a blood clot. It is this blood clot, a "coronary thrombosis," that causes the heart attack.

Having a coronary was first identified in 1878 by the physician Adam Hammer who suspected that the heart of one of his patients had been stopped by a blocked coronary artery. He confirmed this at autopsy, finding that a coronary artery had been clogged by a jelly-like plug of clotted blood. U.S. President

Dwight Eisenhower suffered a coronary while playing golf in 1955 and needed to be hospitalized. The problem was that he was running for reelection in 1956. To downplay the seriousness of the situation, Eisenhower's physicians and staff reported the president had merely had a "mild coronary thrombosis." The president made a point of walking from the car into the hospital without issue and was reportedly doing just fine. He won reelection.

The phrase "don't have a coronary" was first used in the 1960s, meaning calm down or "don't go having a heart attack."

One of the most nerve-racking, and exhilarating, moments in my job is getting called to the emergency department for an acute MI. That patient is scared, and I am scared for the patient. One patient recently asked me, "Am I having a coronary?" Hence this chapter. He had little idea how lucky he was to be speaking with me because half of heart attack deaths occur before people make it to the hospital. My mission is to quickly stabilize patients and get those thrombosed coronaries open as soon as possible—time is muscle!

Chapter Twenty-Six

SEX, RACE, AND ETHNICITY
IN HEART DISEASE

WHEN IT comes to human genetics, we are all remarkably similar—99.9 percent similar. But are some ethnic or racial populations more or less susceptible to heart disease? Is a man more likely to die of a heart attack than a woman? In a TED talk on April 29, 2016, the physicist Riccardo Sabatini noted that the entire human genetic code could fit into 262,000 pages, or 175 large books. Of those books, approximately 500 pages would be unique to each individual. If intelligent life from outer space found us, it is likely that they would see us all as brothers and sisters, even twins. If only we humans could all see it that way.

Race and ethnicity are social constructs and have little biological or genetic basis, but these terms are often used to refer to people sharing a distinctive physical appearance (such as skin color or country of birth) with ancestral origins in Africa, Asia, or Europe. Could genetic factors increase the odds that a certain race or ethnicity is at greater risk for heart disease? Consider these differences. In the United States, African Americans have an earlier onset of high blood pressure and are 30 percent more likely to die from heart disease or stroke than non-Hispanic whites?[1] The majority of Native Americans die of heart disease, and 36 percent of these deaths occur before age sixty-five.[2] Hispanic Americans and Asian Americans have a significantly higher prevalence of diabetes than non-Hispanic whites.

Genetics may contribute a small amount to the risk of heart disease, but the reality is that lifestyle and environment are the major determinants for most of us who will have a heart attack. Heart disease is the leading cause of death for people of most racial and ethnic groups worldwide, but are there potential genetic differences among us humans that affect heart disease risk? Or are other factors at play?

Cardiovascular diseases are the most common (and among the most preventable) causes of death in the Western world.[3] Economic development in Asia, Central and South America, and Africa are resulting in lifestyle changes and environmental exposures that are accompanied by a rapid increase in heart disease deaths in these developing countries. Cardiovascular deaths account for 32 percent of the world's total deaths each year (a 17 percent increase over the past decade).[4] New risk factors such as obesity, stress, and an unhealthy lifestyle are contributing to an increase in heart-related problems worldwide (figure 26.1). Regardless of race, 80 percent of premature heart attacks and strokes are preventable.[5] So at most only about 20 percent of cardiovascular risk could be genetic

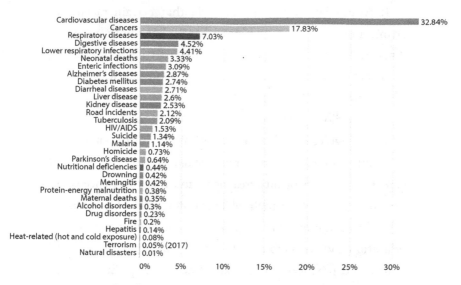

FIGURE 26.1 Leading causes of death globally in 2019.
Source: Courtesy of Our World in Data.

In the United States, cardiovascular disease accounts for approximately one-third of the disparity in potential life-years between Blacks and whites. A genetic difference that predisposes Blacks to high blood pressure might play a role. Some researchers believe that people who lived in equatorial Africa developed a genetic predisposition to being salt sensitive, which means their bodies retain more sodium. This condition increases blood volume, which in turn raises blood pressure. Salt sensitivity allows the body to conserve water, which can be beneficial in a hot, dry climate. Generations later, however, the American descendants of these individuals remain disproportionately salt sensitive. Earlier onset of high blood pressure can lead to premature stroke or heart attack.

It would be simple to conclude that certain racial and ethnic minorities in the United States have both a genetic disposition to and behavioral practices (diet, lack of exercise, etc.) that contribute to premature and increased heart disease risks. But that explanation doesn't account for what is really driving cardiovascular disease disparities. In the United States and worldwide, many racial and ethnic minority populations confront more barriers to cardiovascular disease diagnosis and care, receive lower quality treatment, and thus experience worse health outcomes than their white counterparts.[6] Disparities are related to a number of complex factors including income and education and access to care. Language differences and cultural beliefs and practices can also affect access to quality care and health-seeking behaviors.

Simply put, your ZIP code is more important to heart health than your genetic code. A person's neighborhood and how safe it feels can have an impact on the ability to exercise and eat healthy foods. Low-income neighborhoods can be food deserts, where healthy fresh foods are unavailable and unhealthy fast food is less expensive anyway. A physician telling such a person to eat healthier and go for walks ignores the underlying environmental conditions and related stressors. In addition, the feeling of being unsafe in a neighborhood may cause daily stress and increase stress-related hormones, both known to increase heart disease events. The accumulated weight of the daily lives of minority groups have not been considered in assessing and treating their risk for heart disease.

The majority of differences in heart disease rates, age of onset, and increased prevalence of risk factors, such as diabetes, smoking, and obesity, are related to socioeconomic status, physical environment, employment status, access to care, and social support. Social determinants of health make up a larger part of cardiovascular risk factors than nearly any other area of health. To determine if this disparity gap accounts for the majority of differences in heart disease, we need to first ensure access to quality health care for all. This is a health equity imperative. Culturally sensitive preventive heart care programs and improved food sourcing and safety are necessary for the health of these communities. Increasing the number of people of color in medicine could improve cultural competency among health care providers as well. We need to increase research (and act on it!) on environmental and genetic differences that may contribute to health disparities. Fortunately, the American College of Cardiology and the American Heart Association have taken this, and gender inequality, as central missions.[7]

■ ■ ■

In the United States, men have been more likely than women to die of heart attacks, but as our population ages, women appear to be catching up. Heart disease develops seven to ten years later in women than in men.[9] Estrogen is believed to offer protection from heart disease until after menopause (conversely, this gender difference could be due to the adverse effects of androgens in men). Yet in the United States heart disease is the major cause of death in women, and since 1987 more women than men have died of heart disease each year.[10]

When experiencing a heart attack, men tend to have the classic symptom of chest pressure, whereas women often present with atypical symptoms, such as sudden shortness of breath, bad indigestion, new or dramatic fatigue, or pain in the neck, jaw, or back. This often delays treatment in women, and as a result, women tend not to do as well after a heart attack. Women do not receive comparable treatments, their stay in the hospital is longer, and fewer women survive. Furthermore, women are generally older when they have their first heart attack and are 50 percent more likely to die than men.

Despite the fact that heart disease kills ten times as many women every year as does breast cancer, research suggests that physicians are less likely to discuss heart disease risks with women than with men.[11] Thus, women are less likely to receive guideline-directed preventive care despite the fact that at least two major risk factors for heart disease—diabetes and smoking—increase women's risk of heart attack more than men's.

■ ■ ■

In recent years, people curious about their ancestry have performed at-home DNA tests that analyze the small variations in DNA that determine an individual's racial and ethnic heritage. In the near future, physician-scientists will be able to analyze an individual's unique DNA and understand the genetic predispositions or susceptibilities the individual may have to specific diseases. With that information, we will be better able to determine the extent to which genetic and environmental factors play a role in heart disease risk.

Chapter Twenty-Seven

SUDDEN DEATH OF AN ATHLETE

The problem with heart disease is the first symptom is often fatal.
Michael Phelps, U.S. Olympic swimmer

REMEMBER PISTOL Pete Maravich of the NBA Jazz, Hank Gathers of Loyola Marymount University, Reggie Lewis of the Boston Celtics, and marathoner Ryan Shay? All tragically died during sporting events due to unsuspected heart disease. Approximately seventy-five male and female athletes between the ages of thirteen and twenty-five suddenly die each year in the United States.[1] Worldwide sudden cardiac death during sport activity occurs with an incidence of one in fifty thousand athletes per year.[2] These deaths occur during or immediately after exercise.

Born with unsuspected heart conditions, the hearts of these athletes suddenly fibrillate (life-threatening heart rhythm that results in a rapid, inadequate heartbeat) during exercise, and then stop beating. These athletes collapse and die within minutes if not resuscitated immediately. Congenital heart conditions associated with sudden death of athletes include hypertrophic cardiomyopathy (very thick heart muscle), coronary anomalies (abnormal take off of a coronary artery or its course in the heart muscle), as well as arrhythmogenic right ventricular dysplasia and long QT syndrome (both cause life-threatening arrhythmias).

Screening young athletes may help prevent these tragedies. I am honored to have worked with Simon's Heart, an organization founded by Phyllis and Darren Sudman, who lost their infant son to sudden death because of long QT syndrome. The Simon's Heart team has screened the hearts of thousands of high school athletes, and they have educated even more about athlete sudden death and how to use automatic external defibrillators (AEDs) in their schools. Simon's Heart engages in regional and national legislative initiatives that raise awareness about sudden death in young athletes and save lives.

■ ■ ■

The sudden death of a young athlete is relatively rare, but it is heartbreaking. We must remember that for the vast majority of people exercise does improve heart health. I always tell my patients, "it's a muscle, exercise it." Regular aerobic exercise improves cardiovascular health through multiple mechanisms. These include lower blood pressure, improved cholesterol levels, better sugar regulation, reduced body weight, reduced systemic inflammation, and improved mental well-being. Exercise improves the function of the heart and body arteries and positively modulates the sympathetic nervous system (that heart-brain connection).

How much exercise? Current recommendations by the U.S. Centers for Disease Control and the American Heart Association suggest thirty minutes per day on five days of the week or at least 150 minutes a week of "heart-pumping" exercise.[3] Examples of moderate aerobic exercise are brisk walking, jogging, swimming, and cycling. Playing sports, such as tennis, basketball, and soccer, also count. Regular exercise is as good as any medicine you can be placed on to protect your heart; it will reduce your risk of heart disease by up to 50 percent.

Chapter Twenty-Eight

THE WORD *HEART*

HEART COMES from the Old English *heorte*, which comes from the proto-Germanic **hertan-*, which comes from the proto Indo-European **kerd-*, which derives from the Greek *kardia* and the Latin *cor* or *cord-*. In the Old English, heorte had multiple meanings: breast, soul, spirit, courage, memory, and intellect (to learn "by heart").

From the Latin *cor*, a "cordial" is a medicine to stimulate the heart, and "being cordial" is the notion of warm, friendly feelings emanating from the heart. "Record" originally meant to learn by heart, before it became used for more tangible ways to store information. "Courage" first meant "to speak one's mind by telling all one's heart," long before its meaning became more limited to bravery.

The word *heart* has taken on a plethora of meanings throughout the ages. This comes as no surprise as our ancestors believed that the heart created emotions, gave courage, held memory, and housed the soul.

■ ■ ■

The phrase "absence makes the heart grow fonder" is thought to have come from the first century BCE when the Roman poet Sextus Propertius wrote: "Always toward absent lovers love's tide stronger flows."

A "bleeding heart" originally described someone who expressed great sympathy for another's misfortunes. Used as early as the fourteenth century by Geoffrey Chaucer, bleeding heart became associated with the heart of Jesus and his lamentations on behalf of the poor and sick. However, by the twentieth century, bleeding heart had become a derogatory term for someone who expressed excessive sympathy for another's unfortunate situation. The American journalist Westbrook Pegler employed the phrase, first in 1938, deriding the Roosevelt and Truman administrations. The phrase was later used by Senator Joe McCarthy to attack suspected communists, then by conservative politicians against their "bleeding heart" liberal colleagues. In 2015, Congressman Jack Kemp generalized the term for all to use with his biography subtitled, *The Bleeding-Heart Conservative Who Changed America*. It is also the name of a plant from the genus Dicentra that has heart-shaped flowers that droop.

"From the bottom of one's heart" was first recorded in the *Book of Common Prayer* in 1545: "Be content to forgive from the bottom of the heart all that the other hath trespassed against him." Meaning "most sincerely," it implies deep-seated feelings residing at the deepest part of the heart. The expression may be traced to Virgil's epic poem *Aeneid* (29–19 BCE). For Virgil, the heart was the seat of thinking and feeling. The most profound, deep-seated feelings resided in the deepest parts of the heart. From the *Aeneid*: "Then Aeneas truly heaves a deep sigh, from the depths of his heart." And later, "So his voice utters, and sick with the weight of care, he pretends hope, in his look, and stifles the pain deep in his heart."

■ ■ ■

Shakespeare first used the phrase "to your heart's content" in both *Henry VI*, Part 2 (1591) and *The Merchant of Venice* (1599). Wearing your "heart on your sleeve" was utilized by Shakespeare in *Othello* (1604). The phrase previously had referred to knights wearing ribbons from ladies tied around their arm when jousting to show the one they fought for had their heart.

To "warm the cockles of one's heart" means to bestow contentment on a person and to kindle warm feelings in the person. The phrase dates back to the mid-1600s, a time when scientific texts were often written in Latin. The Latin term *cochleae cordis* means "ventricles of the heart," and the theory goes that the word *cochleae* was corrupted as "cockles." This may have been a mistake or a joke, but it stuck. Another theory is the fact that the cockle, a bivalve mollusk, is shaped like a heart. I would offer one more theory. In the Middle Ages in Europe, one treatment for chest pains was eating cockles cooked in warm milk. So warming the cockles made one's heart happy.

"Cross my heart and hope to die, stick a needle in my eye" first appeared in the late 1800s. This promise attesting to the truth of something originated as a religious oath based on the sign of the cross. Catholic children, as they swore what they said was true, would make an "X" over their heart (the sign of the cross) and then point skyward (to God).

■ ■ ■

The phrase *heart disease* was first used in 1830; *heart attack* in 1836; and *heart beat* in 1850. Other heart idioms were also coined over the years: heartfelt, heartwarming, heavy heart, heart strings, heart's desire, heartache, heartthrob, sweetheart, and heart and soul.

Phrases that feature the heart continue to spread: take it to heart, with all one's heart, to get to the heart of something, to feel a hole in one's heart, have a change of heart, eat your heart out, to have a heart of gold, to have a heart of stone, and to feel in one's heart of hearts. And let's not forget the more recent descriptive meal to end all meals—heart attack on a plate.

PART 5

THE MODERN HEART

Chapter Twenty-Nine

ENLIGHTENMENT AND THE AGE
OF REVOLUTION

There is a wisdom of the head, and a wisdom of the heart.
Charles Dickens, 1854

BY THE END OF the seventeenth century, anatomical knowledge of the heart was surprisingly accurate, and William Harvey's theories of a double circuit, made up of the pulmonary and systemic circulations, became widely accepted. It was during the Renaissance that science changed how we viewed the heart. No longer the de facto seat of emotion and intellect, no longer the very soul, physicians and scientists came to believe that the heart was merely a mechanical pump devoid of spiritual and emotional significance. Enlightenment and a revolution in our understanding of the workings of the heart and circulatory system, the recognition of diseases of the heart, and how to diagnose and treat them, advanced from the mid-seventeenth century through the nineteenth century, coinciding with a revolution in industrial mechanization.

Anatomical exploration by the English physician Thomas Willis led him to assign behavioral and physiological functions to specific parts of the brain in 1664.[1] His theories laid the foundations for the field of neurology and established the brain as the center of intelligence. The brain's primacy over the other organs became entrenched. The heart began to be viewed as little more than a mechanized pump during the Enlightenment (seventeenth and eighteenth centuries) and in the Age of Revolution (eighteenth and nineteenth centuries).

After millennia of speculation, Marcello Malpighi finally established the presence of capillaries connecting the arterial and venous trees in 1661.[2] Born the same year Harvey published his historic work on the pumping heart and blood circulation, the Italian scientist examined arteries and veins in frog lungs with a new device called a microscope. He observed that capillaries connected the smallest arteries, called arterioles, to the smallest veins, or venules. The walls of capillaries were a single cell thick, and they were everywhere. No cell in the body is more than twenty microns (about a third of a hair diameter) away from a capillary.

Richard Lower, credited with being the first to understand that blood circulates through the lungs to become oxygenated, performed the first blood transfusion in 1669 when he attached the arteries of two dogs with a goose feather quill. He later tried it between a "gentle" lamb and a mentally unstable man named Arthur Coga.[3] Coga survived the transfusion and received twenty schillings, which he spent on drink, but his mental illness did not improve. Research on blood transfusion was stymied for another one hundred years.

Raymond de Vieussens, a French anatomy professor, published *Nouvelles Découvertes sur le Coeur* (New Discoveries of the Heart) in 1706, in which he presented a detailed anatomy on the blood vessels of the heart—the coronary arteries and veins.[4] In his 1715 *Traité Nouveau de la Structure et des Causes du Mouvement Naturel du Coeur* (Treatise on the Structure of the Heart and the Causes of Its Natural Motion), Vieussens described in detail the pericardium, the sac holding the heart, and the orientation of the heart's muscle fibers (Galen had observed the three directions of the heart's muscle fibers 1,500 years earlier). Vieussens also described the first clinical presentations and autopsy results of patients with mitral valve stenosis (narrowed heart valve) and aortic valve regurgitation (leaky valve).

The heart and circulatory system were now the domain of physicians and scientists; however, heart diseases were still believed to be extremely rare. In the *Encyclopédie* (1751), Denis Diderot and Jean le Rond d'Alembert wrote, "Generally speaking, we can say that diseases of the heart are rare." Little had changed since Pliny the Elder wrote in the first century CE that "the heart is the only internal organ which disease cannot touch, and which does not prolong the sufferings of life." But this idea began to change in the Age of Revolution. Nineteenth-century physicians started to identify pains in the chest as heart-related and life-ending. Heart pains were becoming more frequent as people were living longer. Life expectancy in the United States before 1800 was less than thirty years, increasing to fifty-four years by 1917—and was seventy-nine years in 2019, for perspective.[5]

Despite this disillusionment of the heart as another organ that can become diseased, the heart as a symbol of love continued to be used in literature, music, and day-to-day life. The first Valentine's Day cards appeared in the 1700s, and they had hearts on them.

■ ■ ■

In 1733, Stephen Hales, an English clergyman and scientist, made measurements of blood pressure in several animal species by inserting thin brass pipes and glass tubes into arteries and measuring the height to which the column of blood rose. In his book *Statical Essays: Containing Haemastaticks, or an Account of Some Hydraulick and Hydrostatical Experiments Made on the Blood and Blood Vessels of Animals* he described the first blood pressure measurement:

> I caused a mare to be tied down alive on her back; she was 14 hands high, and about 14 years of age, had a fistula on her withers, was neither very lean nor very lusty: having laid open the left crural artery about 3 inches from her belly, I inserted into it a brass pipe whose bore was 1/6 of an inch in diameter; and to that, by means of another brass pipe which was fitly adapted to it, I fixed a glass tube, of nearly the same diameter, which was 9 feet in length; then untying the ligature on the artery, the blood rose in the tube 8 feet 3 inches perpendicular above the level of the left ventricle of the heart: but it did not attain to its full height at once; . . . when it was at its full height, it would rise and fall at and after each pulse 2, 3, or 4 inches.

It would be another 163 years before blood pressure could be accurately and routinely measured in humans.

■ ■ ■

Listening to the heart dates back at least as far as the ancient Egyptians. Hippocrates described placing an ear on the patient's chest to discern the sounds of the heart and lungs (direct auscultation).[6] In one case he reported hearing a sound "like the boiling of vinegar" in a dying patient; this is a classic description of what we now know to be acute congestive heart failure. More than a thousand years later, Harvey described the sounds of the heart, also using direct auscultation, as "two clacks of a water bellows to rayse water." The practice of placing an ear to the patient's chest continued until 1816, when René Théophile Hyacinthe Laënnec (1781–1826) invented the stethoscope.[7] Laënnec saw children playing with a log in the courtyard of the Louvre. The children held their ear to one end of the log while the opposite end was scratched with a pin. The log transmitted and amplified the scratching sound. Laënnec, who had a musical ear as a lifelong flutist, was inspired. He wrote in his treatise *De l'Auscultation Médiate* (1819):

> In 1816 I was consulted by a young woman labouring under general symptoms of a diseased heart, and in whose case percussion and the application of the hand were of little avail on account of the great degree of fatness. The other method just mentioned [direct auscultation] being rendered inadmissible by the age and sex of the patient, I happened to recollect a simple and well-known fact in acoustics, . . . the great distinctness with which we hear the scratch of a pin at one end of a piece of wood on applying our ear to the other. Immediately, on this suggestion, I rolled a quire of paper into a kind of cylinder and applied one end of it to the region of

the heart and the other to my ear, and was not a little surprised and pleased to find that I could thereby perceive the
action of the heart in a manner much more clear and distinct
than I had ever been able to do by the immediate application
of my ear.

■ ■ ■

Physicians and scientists in the eighteenth and nineteenth
centuries began to develop treatments for some types of heart
disease, such as dropsy (the swelling of soft tissues, especially
in the legs, due to the accumulation of fluid; what we now call
"edema" due to congestive heart failure). William Withering
(1741–1799), a British physician and botanist, evaluated a folk
remedy for dropsy. The remedy was made up of more than
twenty herbs. He determined that foxglove was the active
ingredient. Using his poor patients to test foxglove, he found
the drug to be helpful in reducing the swelling associated with
dropsy. *An Account of the Foxglove and Some of Its Medical Uses*
(1785) was the first systematic description of how a plant could
be used for therapeutic purposes, in this case for heart failure.

■ ■ ■

By the beginning of the nineteenth century, physicians began
to understand that the heart was fallible and that physical
examination could help diagnose a sick heart. Physician-
scientists began to name diseases associated with the demystified heart: angina pectoris (chest or heart pains), endocarditis

(infection of the heart or its valves), pericarditis (inflammation of the heart sac), and myocardial infarction (heart attack). The term "arteriosclerosis" (from the Greek *arteria* meaning artery and *sklerosis* meaning hardening) was first used in 1833 by the French pathologist Jean Lobstein to describe the condition in which calcified fatty deposits gathered on the inner walls of arteries with age, constricting and hardening them.

William Heberden delivered a paper at London's Royal College of Physicians in 1768 that described a patient who suffered crushing sensations in his chest in response to physical exertion.[8] This pain was alleviated by rest. He called this condition "angina pectoris," from the Greek word for strangling, *ankhone*, and the Latin word for chest, *pectus*. It is worth noting that Heberden first misdiagnosed angina pectoris, thinking it was due to a stomach ulcer and not the heart. He did correctly observe that if it worsened the patient might suddenly lose consciousness and die.

It was not until forty years later that Sir Thomas Lauder Brunton proposed amyl nitrate as a remedy for angina pectoris. Amyl nitrate was also used as a first aid measure for cyanide poisoning, and as an additive in diesel, where it accelerates the ignition of the fuel. It worked well in relieving chest pains— except for the associated pounding headaches. He later tried a related compound, nitroglycerin, finding it even more effective in reducing heart pains. And, yes, nitroglycerin is the active ingredient in dynamite.

Daniel Hale Williams, an African American-Scottish-Irish-Shawnee physician, founded Provident Hospital in Cook County, Illinois, in 1891, making it the nation's first racially integrated facility for Black doctors and nurses. The hospital was championed by Frederick Douglass for giving African Americans another choice for care besides the overcrowded charity hospitals. When Williams made an incision into a stab wound in the chest of a man injured in a fight in 1893, he was able to inspect the heart and sew the pericardium (the sac holding the heart) back together with catgut thread.[9] This marked the birth of heart surgery. Another surgeon, Henry C. Dalton, had performed a similar surgery in Alabama on a stab victim two years earlier, but his work was not published until after Williams's article. Although these were not surgeries on the heart muscle itself, but on the surrounding pericardium, it was the beginning of a new era of surgery on the heart.

The first true surgery on the heart itself took place three years later in 1896, when Ludwig Rehn of Frankfurt Germany stuck his finger into the heart muscle of a twenty-two-year-old gardener who had been stabbed in the heart while walking in a park.

"Digital pressure controlled the bleeding, but my finger tended to slip off of the rapidly beating heart. The heart's contractions were unaffected by my touch." He successfully sewed up the hole in the heart with catgut suture. "The first suture stemmed the flow of blood. Placement of the second suture was greatly facilitated by traction on the first. It was very disquieting to see the heart pause in diastole [stop beating] with each pass of the needle. After the third suture, the bleeding stopped completely. The heart gave a labored beat, and then resumed with forceful contractions as we breathed a sigh of relief."[10]

The patient survived the first cardiorrhaphy, or suturing of the heart muscle, and the field of heart surgery was officially born.

■ ■ ■

The mummified heart of King Louis XIV was stolen during the French Revolution (1789–1799). It eventually ended up in the hands of Lord Harcourt of Nuneham House, in Oxfordshire, England. At a dinner party in 1848, he circulated the walnut sized heart for guests to admire.

William Buckland, the polymath Dean of Westminster, known for his expertise in geology, paleontology, and theology, was one of the guests at the dinner. Buckland pioneered the use of fossilized feces in reconstructing ecosystems, coining the term "coprolites"—fossilized poop. He was known for performing his geological work wearing an academic gown and giving lectures with a dramatic delivery, occasionally on horseback.

William Buckland's home was filled with specimens—animal and mineral, live and dead. His ultimate goal was to taste every animal on Earth, known as zoöphagy. He was known to serve his guests such delicacies as panther, crocodile, and mouse. After being passed King Louis's heart at Lord Harcourt's dinner party, Buckland exclaimed, "I have eaten many strange things, but have never eaten the heart of a king before." Before anyone could hold him back, he popped it in his mouth.[11]

■ ■ ■

Maid of Athens, ere we part,
Give, oh give me back my heart!
Or, since that has left my breast,
Keep it now, and take the rest!
Hear my vow before I go,
Ζωή μου, σᾶς ἀγαπῶ.

Lord Byron, 1810

Lord Byron's close friend Percy Bysshe Shelley (*Ozymandias* and *Ode to the West Wind*) was just twenty-nine when he drowned. His boat, *Don Juan* (named after Byron's poem), was caught in a storm in 1822. His body was found ten days later and identified by the book of John Keats poems in his pocket. During his pyre-style cremation on the beach, his heart refused to burn (one theory being that the pericardium surrounding the heart had calcified due to an earlier bout of tuberculosis). His friend Edward Trelawny removed it from the pyre and gave it to Mary, the late Shelley's wife and author of *Frankenstein*. She carried it with her in a silken shroud until her death.[12] After her death, her husband's heart was found wrapped in pages from one of his last poems, *Adonais*. The heart remained with the family until it was buried with their son Percy Florence Shelley in 1889. The headstone of Percy Bysshe Shelley reads *Cor Cordium* or "Heart of Hearts."

Chapter Thirty

THE TWENTIETH CENTURY
AND HEART DISEASE

*Occasionally in life there are those moments of unutterable
fulfillment which cannot be completely explained by those symbols
called words. Their meanings can only be articulated by the inau-
dible language of the heart.*

Martin Luther King Jr.

*The prudence of the best heads is often defeated by the tenderness
of the best hearts.*

Winston Churchill

INTO THE twentieth century, the heart remained the metaphor
of choice for our emotional and spiritual lives. Nevertheless,
science and medicine now firmly located our thoughts, passions,
and reasoning in our brain. By 1871 Charles Darwin was calling
the brain "the most important of all organs," but we could still
have a broken heart, metaphorically. We still wore our heart
on our sleeve. On those tough life decisions, we followed our
heart. But if a heart could be transplanted from one person to
another, then the soul could not possibly be in the heart. Our
evolving understanding of the body, the brain, and the heart
seemingly changed forever our conception of our heart; it was
physically just a pump—an important pump—but not the seat
of our emotions, conscience, intelligence, and memory.

Pneumonia was the leading cause of death in the United States in 1900. Heart disease was fourth, behind tuberculosis and diarrhea. But by 1909, and continuing to the present day, heart disease was the leading cause of death in Americans (except between 1918 and 1920 due to the Spanish Flu pandemic).[1] Improvements in sanitation, public health, and medical treatments (such as antibiotics) led to declines in infectious disease deaths. At the same time, life expectancy increased. Chronic diseases, such as heart disease and cancer, became the leading killers. Driven in part by increased smoking rates in the United States (<5 percent in 1900 to 42 percent in 1965), as well as more processed foods, saturated fats, and decreased physical exercise with higher car use, heart disease deaths rose steadily, peaking in the 1950s and 1960s.[2]

The United States Congress passed the National Heart Act in 1948, declaring "the nation's health [to be] seriously threatened by diseases of the heart and circulation." Signing the bill into law, U.S. President Harry Truman called heart disease "our most challenging public health problem." The act created the National Heart Institute (now the National Heart, Lung, and Blood Institute), which is part of the National Institutes of Health.

Public education campaigns helped decrease smoking rates and increased people's awareness of the consequences of high blood pressure and high cholesterol. Physicians and scientists developed more effective treatments for heart disease. Between 1958 and 2010, heart disease death rates have decreased from their peak in the 1950s and 1960s (figure 30.1).[3] However, heart disease still remains the leading killer in the

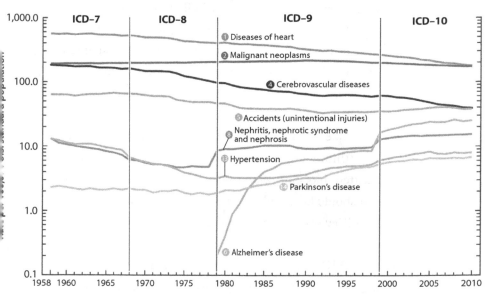

FIGURE 30.1 Age-adjusted death rates for selected leading causes of death: United States, 1958–2010. ICD is the International Classification of Diseases. Circled numbers indicate ranking of conditions as leading causes of death in 2010.

Source: CDC/NCHS, National Vital Statistics, Mortality.

United States and worldwide. For perspective, as of July 2022, 6.3 million people have died worldwide from COVID-19, whereas around 18 million died from cardiovascular disease in the year 2021 alone.

■ ■ ■

Our understanding of heart disease and how to fix it took giant leaps in the twentieth century.[4] In 1899, a Chicago pathologist,

Ludvig Hektoen, suggested that a buildup of atherosclerotic plaque in the coronary arteries caused heart attacks. In 1912, his colleague, physician James B. Herrick, wrote the landmark article, "Clinical Features of Sudden Obstruction of the Coronary Arteries," proposing that a heart attack results not from a coronary artery eventually closing down from too much atherosclerotic plaque building up inside the artery but rather from a blood clot (thrombosis) forming in the atherosclerotic coronary artery, acutely blocking blood flow to the downstream heart muscle. It should be noted that this discovery occurred simultaneously in Kiev by Obrastzow and Straschesko.[5]

Despite this major insight into the cause of a myocardial infarction (MI), the hypothesis was ignored by the medical establishment for almost seventy years. In 1980, the Spokane physician Marcus DeWood published a report on 322 of his heart attack patients. DeWood placed catheters into their coronary arteries within twenty-four hours of their MI, injecting contrast dye while taking X-rays. He showed that blood clots caused the heart attacks. Unfortunately, treatment for a heart attack at that time was morphine, bed rest, and praying. If the clot did not dissolve, the heart muscle beyond the blood clot died (infarcted). Balloon angioplasty had only recently been invented and was not in widespread use yet. Clot busting (thrombolytic) medicines did not come into use until the 1990s.

The German cardiologist Andreas Gruentzig performed the first balloon angioplasty of a blocked coronary artery in 1977 using his kitchen-built balloon Krazy Glued onto the tip of a coronary catheter. The word *angioplasty* comes from the Greek words *angeion* or "vessel" and *plastos* or "shaped or formed." Balloon angioplasty rapidly gained popularity as a nonsurgical alternative to coronary artery bypass graft surgery (CABG; see chapter 32), especially after the addition of stents (tiny metallic chicken-wire-like tubes to keep the ballooned artery from recoiling back shut). By 1990, balloon angioplasty and stenting had become more common than CABG. In the early years of the twenty-first century, stents began to be coated with drugs that prevented scar tissue from forming in the artery. The first coating drug was an antibiotic called Rapamycin, which stops cell division; it was discovered in a soil mold on Easter Island.

Before Gruentzig could open a narrowed coronary artery with a balloon glued onto his catheter, someone had to dare to place the first catheter into a human heart. Werner Forssmann, a German physician, took on the task and performed the first human heart catheterization . . . on himself. The technique was developed in 1844 by Claude Bernard, a French physiologist, who used catheters to record heart pressures in animals and coined the term "cardiac catheterization." In 1929, against the advice of colleagues, Forssmann convinced Gerda Ditzen, an operating room nurse, to get the sterile supplies and assist him. She agreed only if he would perform the procedure on her. He agreed, restrained her on the operating table, then quickly anesthetized his own arm and inserted a urinary catheter into his antecubital vein—the large vein in the crevice in front of the elbow—and pushed it toward his heart.

He then let her up and they walked to the X-ray department, where he had to push the catheter further until it reached his right atrium. He then took the confirmatory X-ray. Instead of accolades, Frossmann was chastised and ostracized from cardiology and subsequently went into urology. He was finally awarded the Nobel prize twenty-seven years later in 1956, after Andre Cournand and Dickinson Richards, physician-scientists working in New York, found his obscure, ridiculed article on the subject and advanced cardiac catheterization for the purposes of recording heart pressures, blood flow, and images of the heart chambers.

Now a catheter could be placed in the heart, contrast injected, and pictures of the heart chambers imaged. But what about imaging the coronary arteries? Frank Mason Sones at the Cleveland Clinic performed the first human coronary angiogram . . . by accident. Sones was performing a heart catheterization in 1958 on a twenty-six-year-old man with rheumatic heart valve disease. He was preparing to take an image of the patient's left ventricle when the catheter accidentally jumped into the right coronary artery as contrast dye was injected. "We've killed him!" Sones exclaimed. The patient's heart stopped, but restarted after repeated deep coughs. The era of coronary angiography was born.

With subsequent modifications of catheters, X-rays, and contrast dyes, coronary angiography became safer, and its use spread rapidly around the world (figure 30.2). Gruentzig took the next step to open blocked coronary arteries using his catheters with balloons glued on them. When you think about it, he was working as a glorified plumber. The heart had lost its

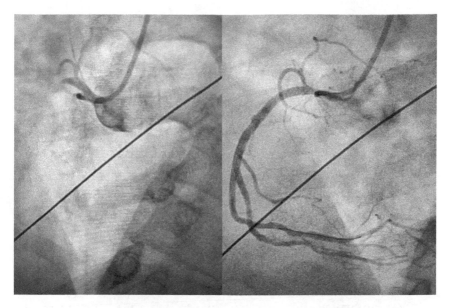

FIGURE 30.2 Coronary angiography before and after balloon angioplasty of the right coronary artery during a heart attack.
Source: Photos by author.

mystique. Now just a machine, doctors could fix its clogged
pipes and get the pump working again. Today, more than a
million cardiac catheterizations are performed yearly. And if
there is any coronary artery disease found on "cath," that patient
gets started on "an aspirin a day."

Chapter Thirty-One

ASPIRIN

Most things in this world don't work, aspirin do.
Kurt Vonnegut

BOTH CUNEIFORM tablets (3500 BCE) and the Ebers Papyrus (1550 BCE) documented that willow and myrtle leaves, which contain salicylic acid—*salix* is Latin for willow—were used by ancient Sumerians and Egyptians to treat aches and pains. Hippocrates (400 BCE) used tea made from willow bark to reduce fevers. Touted by Galen and also used by Chinese, Native American, and African societies, willow bark was used through the Middle Ages and into the nineteenth century as a pain and fever reliever.[1]

Charles Frederic Gerhardt first synthesized acetylsalicylic acid in 1853. Bayer, a drug and dye firm, branded acetylsalicylic acid "Aspirin" and began selling it around the world in 1899. The letter "A" stood for acetyl, "spir" for the Spiraea ulmaria flower (meadowsweet; another natural source of salicin), and "in" was a common suffix used for drugs at the time. Ironically, Bayer's first advertisements stated that aspirin "did not affect the heart."

Fifty years later, the U.S. physician Lawrence Craven observed that four hundred of his male patients who took aspirin for two years had no heart attacks.[2] By 1956, he had recorded eight thousand patients taking aspirin and found no heart attacks in this group. Beginning in 1974, large aspirin trials demonstrated that it prevented heart attacks and death.

How does aspirin prevent heart attacks? Aspirin interferes with one of the blood's clotting mechanisms, and heart attacks are caused by blood clots forming when the atherosclerotic plaques in the wall of diseased coronary arteries crack open. So if you think you are having a heart attack, call emergency medical services first, then chew an aspirin!

Chapter Thirty-Two

THE TWENTIETH CENTURY
AND HEART SURGERY

For a dying man [a heart transplant] is not a difficult decision. . . .
If a lion chases you to the bank of a river filled with crocodiles,
you will leap into the water; convinced you have a chance to
swim to the other side.

Christiaan Barnard, surgeon who performed the first heart transplant

IN 1896, Ludwig Rehn had stuck his finger into the hole of a heart stabbed with a knife and sewed it shut with catgut. After that remarkable feat, there were few advances in the field of heart surgery for fifty years. This was mostly due to the lack of survival because of infection after the surgery. Progress began in 1944 when Helen Taussig—a woman pioneer in cardiology credited with being the founder of pediatric cardiology—worked with the surgeon Alfred Blalock and surgical technician Vivien Thomas (an African American pioneer in cardiology who developed the surgical technique in the animal laboratory and instructed Blalock through the first surgeries on children) to save "blue babies," children with fatal heart defects. Their operations on babies with congenital heart defects caused a sensation. These operations were not "in" the heart itself but were on the large arteries and veins going to and coming from the heart. Taussig, Thomas, and Blalock saved these blue babies and sparked the beginning of the era of modern heart surgery.[1]

One day in 1940, the physician Wilfred Bigelow observed a markedly slowed heart rate in a man who was brought into his office with frostbite. It gave him an idea. He performed experiments on dogs by cooling them and stopping the blood flow to the heart for fifteen minutes. (Remember the six-minute problem? The brain and vital organs can be irreversibly damaged within six minutes when deprived of oxygen-rich blood.) Over half of the dogs recovered.

Based on the research of Bigelow and others, John Lewis, with the assistance of C. Walton Lillehei, performed the first successful "open-heart" surgery using hypothermia in 1952. They sutured a hole in the heart between the left and right atria (atrial septal defect) of a five-year-old girl. She survived.

To save the life of a doomed child, Lillehei (known as the Father of Open-Heart Surgery) stitched the circulatory systems of the one-year-old boy with a hole between his left and right ventricles (ventricular septal defect) to that of his father's (who had his son's same blood type). The procedure effectively turned his father into a heart-lung machine. Lillehei's ideas were inspired by the circulation of blood between mother and fetus. In his early experiments, the circulatory systems of two anesthetized dogs were connected through beer hoses to a milk pump between them that pushed equal amounts of blood in opposite directions without introducing air bubbles. In 1954, Lillehei used the same beer hoses and milk pump during the successful operation between son and father.

At about the same time, John Heysham Gibbon used the first heart-lung machine, which he developed with IBM engineers, to successfully close a large atrial septal defect in an eighteen-year-old woman. The bulky machine temporarily took over the functions of the heart and lungs, removing depleted blood from the body and pumping in oxygenated blood. This allowed Gibbon half an hour to successfully repair the large hole in her heart.

It is interesting that surgeries on the heart became known as open-heart surgeries. Surgeons were thinking purely mechanistically. Metaphorically we think of a person with an open heart as one who is willing to share their deepest thoughts, secrets, and emotions. Now we could both physically and metaphorically have an open heart, but the meanings could not be more different. One referred to the deepest parts of a muscle pump, the other to the deepest parts of our soul.

Cardiac surgeon Albert Starr and engineer Lowell Edwards (also inventor of the hydraulic tree debarker) developed the Starr-Edwards valve in 1960, the first artificial device to be implanted into a heart. A mechanical valve, it was nothing more than a plastic ball that could move back and forth in a cage with the blood flow—and it worked. This was soon followed by Olov Bjork's tilting disk (or toilet seat) valve in the 1970s. Bioprosthetic valves made of pig heart valve or calf pericardial tissue were in use by the late 1960s. And now we can replace heart valves without surgery, putting them in place using a catheter, and the patient goes home the next day!

Rene Favaloro, an Argentine surgeon working at the Cleveland Clinic in 1967, was a pioneer in heart bypass surgery. He successfully grafted a piece of healthy vein from a patient's leg above and below an occluded section of one of his coronary arteries, effectively "bypassing" the blockage; hence the name coronary artery bypass surgery or CABG (said out loud as "cabbage"). David Letterman, Burt Reynolds, and Bill Clinton are all part of the "zipper club," as the surgery is described, because of the scar down the middle of the chest. Or as David Letterman summed it up for Regis Philbin's pending surgery, "they're going to split him open like a lobster." Currently, more than 800,000 CABGs are performed yearly worldwide.

On December 3, 1967, in Cape Town, South Africa, Christiaan Barnard removed the healthy heart of a twenty-five-year-old woman who died in a car accident and transplanted it into the chest of fifty-five-year-old Louis Washkansky, who was dying of heart failure. After the five-hour operation, the transplanted heart was electrically shocked to kick start it. It worked. After waking, Washkansky was able to talk, and he walked soon after. Initially successful, Washkansky died eighteen days later due to pneumonia. The operation was reported in newspapers around the world, making an instant star out of Barnard, who was soon dating Sophia Loren.

Unfortunately, pre-1970 the majority of heart transplants ended in disaster because the patients' bodies rejected the new organs. Barnard was more interested in being the first to perform a heart transplant rather than being the best. Thanks largely to the work of Norman Shumway (who was a third-year surgery resident assisting Lillehei in 1954 and the teacher of Christiaan Barnard), surgeons learned to minimize transplant rejection. Shumway realized blood type mattered. Also, a new drug, cyclosporine, isolated from a fungus found in the soil of the Norwegian forest, prevented organ rejection without damaging the immune system. Currently, more than eight thousand heart transplants are performed yearly worldwide.[2] Now the only problem with heart transplants is that there are not enough donor hearts. The longest living heart transplant recipient has survived with a donor heart for more than thirty-five years.

We now believe that our emotions, our memory, and our thoughts are in our brains, so it's okay to place the heart of one person into another person. It works without a hitch in most cases. But as I wrote in the introduction, along comes someone like Claire Sylvia, the forty-seven-year-old former professional dancer who underwent a heart-lung transplant, receiving a heart taken from an eighteen-year-old man who died in a motorcycle accident. She soon adopted many of the young man's behaviors, which were confirmed by his family. Other accounts of inheriting the personality traits of a transplant patient's donor have been described in the literature. This raises the question of whether we should look at the heart as just a pump. Does it contain pieces of us—whether we want to call that our soul or our emotions—that travel with it? It does so at a minimum symbolically for some of us. A recent news story reported on a bride whose father had passed away and donated his heart. The man who received his transplanted heart walked her down the aisle.

Further challenging ethical and religious senses, the surgeon Leonard Bailey transplanted the heart of a baboon into a twelve-day-old girl, known as Baby Fae, in 1984. Baby Fae had a fatal congenital heart defect called hypoplastic left heart syndrome. She survived with her baboon heart for three weeks. As early as 1964, a Mississippi surgeon named William Hardy transplanted a chimp heart into a dying man that beat for ninety minutes. Attempts with pig and sheep hearts have also proven unsuccessful to date.

Research into xenotransplantation—transplanting organs between members of different species—continues, especially with pigs, the species that has shown the most potential for donor hearts to humans. The idea of another person's heart being placed in our body may already test our faith in individuality, but the idea of another animal's heart within our body raises very different questions. But then, as George Orwell wrote in *Animal Farm* (1946), "The creatures looked from pig to man, and from man to pig, and from pig to man again, but already it was impossible to say which was which."

■ ■ ■

I knew my father was going to die of heart disease, and I was trying to make a heart for him.

Robert Jarvick, inventor of the first artificial heart

Although eight thousand patients receive a heart transplant per year, perhaps ten times that number would benefit from a transplant if a heart was available. There are simply not enough donor hearts. Therefore, replacing the human heart with a mechanical device has been a great ambition of physician-scientists for the last half century. Think of it as the reverse of the Tin Man concept from *The Wizard of Oz*. But like xenotransplantation, will a person with a robot heart "have a heart"?

The first total artificial heart was implanted by Denton Cooley in 1969. It was a temporary bridge to transplantation and was removed three days later. It should be noted that the total artificial heart was being developed in the lab of another heart surgeon, Michael DeBakey. Cooley convinced one of DeBakey's assistants to give him one of the artificial hearts so he could be the first to implant one. "I look upon the heart only as a pump, a servant of the brain," he told *Life Magazine*. "Once the brain is gone, the heart becomes unemployed. Then we must find it other employment." This quote makes clear where the heart stood in comparison to the brain in the minds of physicians and scientists in the twentieth century. The brain defined what is me. The heart was nothing more than a replaceable pump.

■ ■ ■

Willem DeVries implanted the first "permanent" artificial heart, designed by Robert Jarvik, into a retired dentist named Barney Clark in 1982. Clark, who was suffering from severe congestive heart failure, was not a candidate for a heart transplant because of his age and severe emphysema. After Clark received the mechanical heart, his wife asked the doctors, "Will he still be able to love me?" Clark lived for one hundred twelve days tethered to a four-hundred-pound external pneumatic compressor (essentially, an air-driven mechanical pump). Bill Schroeder became the second recipient, surviving 620 days. Research continues on artificial hearts. Current prototypes include soft artificial hearts, which are created with silicone using 3D printing technology. The longest a mechanical heart transplant recipient has lived so far is about five years.

Although artificial hearts that completely replace a broken heart remain limited (less than two thousand implanted worldwide as of 2021), small pumps that can help a failing heart, called ventricular assist devices (VADs), are now routinely implanted. The first VAD was implanted by Michael DeBakey in 1966 into a thirty-seven-year-old woman for ten days until her successful heart transplantation. VADs are supporting devices implanted next to the heart ventricle to pump for the heart muscle, which still may be functioning although poorly. These can be used as a bridge to transplantation (as in the first use by DeBakey), a temporary assist until the patient's own heart recovers, or as "destination" therapy or permanent solution if the person is not a transplant candidate. VADs have become a lifesaving option for end stage heart failure patients. One VAD recipient has lived for over fourteen years.

Medical and technological advances in the twentieth century elucidated the mechanisms underlying heart attacks and heart failure. A heart attack was no longer a death sentence for many; instead it was a mere setback. Interventional cardiologists could open blocked coronary arteries, saving heart muscle. Cardiac surgeons could bypass multiple blocked coronaries, replace damaged valves, and even put in a whole new heart. Yet heart disease remains the number one killer worldwide. What more needs to be done to change this?

Chapter Thirty-Three

THE HEART NOW

Remembering that you are going to die is the best way I know to avoid the trap of thinking you have something to lose. You are already naked. There is no reason not to follow your heart.

Steve Jobs

I will never have a heart attack. I give them.

George Steinbrenner, who died of a heart attack at age eighty

ALTHOUGH WE may not believe the heart is the seat of our emotions today, we continue to subscribe to the heart's symbolic connotations. When walking through most parks, it is common to see tree trunks with a heart carved into them by lovers. The heart appears every Valentine's Day and also on love letters, as emojis, and on my daughter's signature.

We may not believe that the heart is the location of our soul, but we need it to live. One of every three of us worldwide will die from cardiovascular diseases, and cardiovascular diseases kill more of us than all cancers combined. In the United States, someone dies of a heart attack every thirty-six seconds, with 700,000 cardiovascular deaths every year at a cost of $363 billion. In children, the most common congenital diseases are those of the heart.[1]

Recognizing these facts placed cardiology at the forefront of innovation in the twentieth century, and now it is even more so in the twenty first century. The twentieth century saw the development of coronary angiograms, coronary artery bypass surgery, catheter-based coronary balloon angioplasty and stents, pacemakers and defibrillators, heart assist devices, heart transplants, and mechanical artificial hearts. Preventive health measures directed at cardiac risk factors such as smoking, high blood pressure, and cholesterol—and half of Americans currently have at least one of these cardiac risk factors—have helped to decrease deaths due to heart disease. The incidence of cardiovascular disease has decreased significantly since the 1960s, but it remains the number one killer of us all.

■ ■ ■

British scientists say they have developed a super broccoli that can help fight heart disease. You know, if you want to fight heart disease, why don't you come up with a food that people will actually eat? Like a super glazed doughnut.

Jay Leno

The heart symbol has taken on yet another meaning in recent times—it is a sign of health. I only have to look into the bowl of Cheerios that I grab on the fly at work (figure 33.1). The heart-shaped whole grain oats tell me I am eating "heart healthy." When you look at a restaurant menu, what symbol shows you the healthy options?

When you board a plane or are walking down a school hallway, it's hard not to notice the heart symbol crossed with a lightning bolt (figure 33.2). We have come to learn that this tells us an "automatic external defibrillator (AED) is here."

FIGURE 33.1 My "heart healthy" breakfast.

Source: Photo by author.

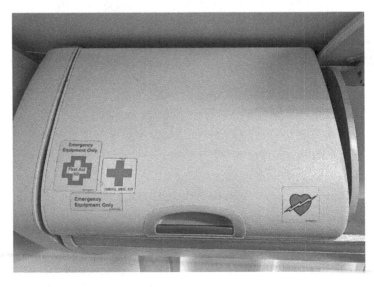

FIGURE 33.2 Heart with a lightning bolt signals an automatic external defibrillator (AED) on board.

Source: Photo by author.

Cutout hearts fill the window of the Palliative Care office at my hospital (figure 33.3). The multicolored hearts seem to be floating up and out the window. Here the hearts symbolize health, hope, gratitude, and love.

FIGURE 33.3 Window of a palliative care hospital office.
Source: Photo by author.

If you fly Southwest Airlines, you may have seen the heart-shaped symbol with the Southwest stripes—the "heart of the people of Southwest Airlines" (figure 33.4). Their literature says this represents the Servant's Heart, rooted in their core value of living by the Golden Rule.

Modern medicine has convinced us that the heart is just another organ in the body, devoid of feelings and reason, and the one that most frequently leads to our demise. Although

FIGURE 33.4 Service with a heart at Southwest Airlines.
Source: Photo by author.

stripped of its importance as the ruler of the other organs, as believed by many ancient cultures, the symbolical power of the heart lives on in modern times. The heart symbol still stands for love and romance, and it has taken on additional value representing health, life, devotion, and service.

We accept that we can metaphorically have our heart broken but dismiss the fact that sudden, profound emotions can really "break" a heart. Emotions may be more closely associated with the heart than we currently accept, and studies suggest that the heart may have more to say about the health of our bodies than we think. Current data seem to indicate a significant back and forth communication between the brain and heart—a heart-brain connection.

Chapter Thirty-Four

BROKEN HEART SYNDROME

The Lord is near to the brokenhearted and saves the crushed in spirit.

Hebrew Bible and Holy Bible, Psalm 34:18

You have to keep breaking your heart, until it opens.

Rumi (1207–1273)

The heart was made to be broken.

Oscar Wilde (1854–1900)

Blessed are the hearts that can bend; they shall never be broken.

Albert Camus (1913–1960)

Hearts will never be practical until they can be made unbreakable.

Tin Man in The Wizard of Oz

What is stronger than the human heart which shatters over and over, and still lives?

Rupi Kaur (b. 1992)

THESE QUOTES, dating back three thousand years, demonstrate the repeated use throughout human history of the metaphoric broken heart in literature, philosophy, and religion. Even today the concept of breaking one's heart remains one of the most known and used metaphors. But can strong emotions truly break a heart? Well, yes, they can.

■ ■ ■

If a tiger chases you, a number of changes will happen in your body; this is referred to as the fight or flight response. The amygdala in your brain triggers a signal telling your body to run. That signal travels to the adrenal glands, where the adrenal medullae release adrenaline. The adrenaline quickly makes its way to your heart's pacemaker cells and speeds up your heart rate. It also causes heart muscle cells to let in more calcium, which makes them contract harder. This provides oxygenated blood to your leg muscles so you can run. However, on occasion, the body's stress system can run amok and damage the heart, causing a stress-induced heart attack.

Stress-related heart attacks are also referred to as "broken heart" or *takotsubo* syndrome. Takotsubo is Japanese for "octopus pot." First described in 1990, Japanese physicians were observing patients, mostly women, who suffered a heart attack after experiencing acute severe emotional grief or stress.[1] Their heart dysfunction made their left ventricle look like an octopus pot—a pot with a wide bottom and a narrow neck (figure 34.1).[2] These patients had classic signs and symptoms of a heart attack: chest pain, elevated heart enzymes, EKG changes, and regional heart wall motion abnormalities. But on cardiac catheterization their coronary arteries were found to be free of atherosclerotic disease.

In the majority of broken heart syndrome cases, heart function recovers. We know that the resulting abnormal shape

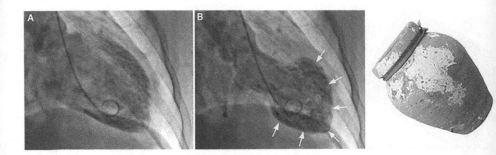

FIGURE 34.1 Left ventriculography in a patient with Takotsubo syndrome. On the left, an end-diastolic image. On the right, an end-systolic image showing hyperkinesis of the base of the left ventricle (dark arrows) but akinesis of the mid and apical segments (light arrows). The shape resembles a traditional Japanese octopus trap (takotsubo), shown on the far right.

Source: With permission from R. Diaz-Navarro, *British Journal of Cardiology* 28 (2021): 30–34.

of the Takotsubo heart reflects the distribution of adrenaline receptors in the normal heart muscle, but we don't know exactly why stress-induced heart attacks happen. A sudden surge of adrenaline can damage heart cells.[3] Studies after earthquakes—1994 in Northridge, California, and 1995 in Kobe, Japan—found that the incidence of heart attacks was much higher on the day of the earthquake than on the same day the previous year.[4] In addition, during and immediately after World Cup penalty shoot outs and Super Bowls, a surge in stress-induced heart attacks occurs.

Sudden severe emotions, or acute stress, can cause a heart to literally break. Fortunately, in many cases that broken heart recovers and the patient survives. As the seventeenth-century

English poet Lord Byron wrote, "The heart will break, but broken live on." Under no other conditions do the metaphorical
heart and the biological heart intersect more closely.

■ ■ ■

Are we surprised that lifelong couples tend to die within months
of each other? Johnny Cash and June Carter Cash died within
four months of each other. The theory is that death of the surviving spouse follows quickly during bereavement because of
the intense physical stress of grief—and a broken heart.[5]

The single saddest moment of my career occurred when
I was a cardiology fellow. Stepping out of a patient's room,
I had to inform her husband of over sixty years that his wife
had passed away. He knew, as well as we did, that she was not
going to survive. But as I told him the news, I watched his face
melt into such anguish and fear. He looked up at me and asked,
"What do I do without her?" The sorrow in his eyes hurts me to
this day. He grabbed my shoulders for support and kept looking at me for an answer. I held this small man in my arms and
cried with him for a long time that day. He passed away five
months later in hospice care.

Despite all of our science and the disillusionment of the
heart as nothing more than a pump, these cases seem to describe
moments when the emotional and physiological parts of our
heart become one. The sixteenth-century anatomist Gabriele
Falloppio stated, "Man cannot live with a broken heart."

The emotions we feel in our brain reverberate in our heart,
and the resulting physical sensations are manifestations of the

heart's response. This interdependence, the heart-brain connection, is vital to our health. It's what led humans over thousands of years to place our emotions, reasoning, and very soul in this hot, pumping organ that signifies we're alive. Ancient societies taught that a happy heart meant a happy body and a long, healthy life. Modern science and medicine are now suggesting that our ancestors may have been more insightful than once thought. It may be that the heart's role in our emotional and physical well-being is more significant than what physicians and scientist have led us to believe in the past five hundred years. It may be that the heart talks to the brain as much as the brain directs the heart, and that this heart-brain connection plays a vital role in our overall health.

Chapter Thirty-Five

THE HEART-BRAIN CONNECTION

*Still I tried to picture a moment when the beating of my heart
no longer echoed in my head.*
Albert Camus

THE SAME ancient letter or symbol in many Asian languages can
mean either heart or mind. Ancient cultures believed the two
were connected, and recent studies suggest that our ancestors
were not so wrong after all. Modern medicine now places the
mind in the brain, but scientists are now proving that the concept of the heart-brain connection is real—and physical.

We know that smoking, high blood pressure, high cholesterol, and diabetes are major risk factors for heart disease. We
pay attention to and try to treat these traditional risk factors to
reduce heart attacks and heart failure. But recent studies suggest
that we are ignoring another major risk factor for heart disease—
emotional stress. Despite the millennia-old association of the
heart with emotions, we have seemingly forgotten the potential
effects of emotion on the health of our hearts.

A growing body of evidence supports a relationship between
psychosocial or mental stress (including depression, anxiety,

anger/hostility) and chronic disease progression (such as heart disease and cancer).[1] We know that acute stressors, such as an earthquake or the sudden loss of a loved one, cause heart attacks. We are now learning that chronic stressors—work stress, marriage stress, or financial stress—can be associated with increased cardiovascular events.

Chronic stress can lead to negative behaviors such as smoking, alcohol bingeing and abuse, poor dietary compliance, physical inactivity, and poor adherence to medical regimens, all of which negatively affect the heart. But chronic stress also leads to negative effects on the sympathetic nervous system, increased cortisol levels, and inflammation and abnormal function of blood vessels, all known mediators of cardiovascular disease. We now know that psychosocial and mental stress can be both a cause and a consequence of heart attacks.

The global INTERHEART Study investigated the association between chronic stressors and the incidence of heart attacks in nearly 25,000 people from fifty-two countries.[2] After adjusting for age, gender, geographic region, and smoking, those who reported "permanent stress" at work or at home had greater than 2.1 times the risk for having a heart attack. Data are now showing that stress management can reduce future cardiac events. Techniques aimed at improving positive emotions, such as yoga, meditation, music, and laughter, can reverse the negative effects of chronic stress on the body.[3] They can also reduce blood pressure and depression. Unfortunately, the influence of psychosocial and mental stressors on the heart is often not recognized alongside traditional cardiac risk factors.

■ ■ ■

Until the 1990s we were taught that the brain unilaterally issued commands to the heart. A new area of research, neurocardiology, is now finding that a dynamic two-way dialogue between the heart and the brain continuously affects the functions of both organs.[4] The heart has an intrinsic nervous system composed of more than 40,000 sensory neurons; it is a "little brain" that enables the heart to sense, regulate, and remember.[5] The heart sends at least as many nerve signals to the brain via the vagus nerve as the brain sends to the heart. Signals from the heart's intrinsic nervous system affect functions in multiple parts of the brain associated with emotions, including the medulla, hypothalamus, thalamus, cerebral cortex, and the emotion center in the brain, the amygdala. Scientists from Thomas Jefferson University in Philadelphia recently used knife-edge scanning microscopy to create 3D models of rat hearts (figure 35.1). They demonstrated visually that the heart has a "little brain," the intracardiac nervous system.[6]

The heart can also affect the brain through hormone and neurotransmitter release. Concentrations of oxytocin, the "love hormone," produced in the heart are in the same range as the amounts made in the brain. Oxytocin affects cognition, tolerance, trust, friendship, and bonding. The heart may also affect the brain through rhythmic electromagnetic energy.[7] The heart is the most powerful generator of electromagnetic energy (remember it has its own electrical system) in the body; it is sixty times greater than the brain.

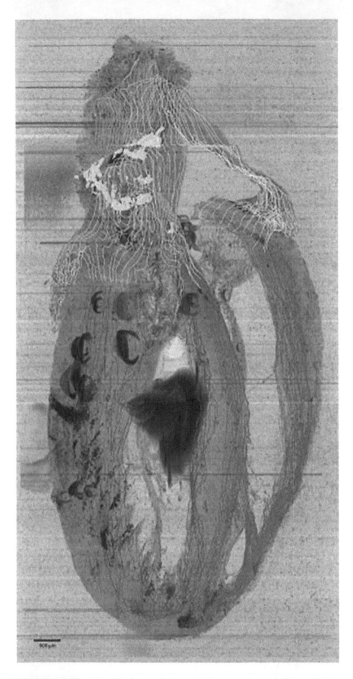

FIGURE 35.1 Nerve cells (white) that make up a heart's "brain" cluster near the top of this rat heart slice, near where blood vessels enter and exit the heart.

Source: With permission from S. Achanta et al., *iScience* 23, no. 6 (June 2020):101140.

A negative example of the effects of the heart on the brain can be seen in many individuals with panic disorder. Research suggests that psychological aspects of panic disorder are often created by unrecognized heart arrhythmias. The sudden marked change in the pattern of signals coming from the heart to the brain, relative to the usual stable baseline rhythmic pattern, can result in anxiety and panic. In many cases, diagnosing and treating the heart arrhythmias ameliorates panic disorder symptoms.

For people with performance anxiety, a common method to calm the nerves is to take medications called beta blockers. These drugs block the effects of adrenaline (which increases heart rate and blood pressure) on the heart. The brain anticipates being anxious immediately before a performance. But when the heart is sending the medication-induced signal that there is no anxiety response, the brain accepts the command from the heart, overruling its need to be anxious.

Coherence methods such as meditation or mindfulness, which modulate heart rhythm and function, can synchronize with other body systems (such as breathing and blood pressure) to positively affect pain regions in the brain. Positive emotions, such as compassion or appreciation, can make the heart's rhythm more coherent and harmonious. This information is sent to the brain, improving one's "state of mind." Thus, the heart's rhythmic beating patterns not only reflect one's emotional state but also play a role in determining our emotional experience. Research suggests that the pattern and stability of the heart's rhythm influences higher brain centers, affecting psychological factors such as attention, motivation, pain perception, and emotional processing.

In addition, data suggest that our hearts can entrain the hearts of others around us. Music is one proven method through which people can improve their emotional state, so it is not surprising to learn that all of the singers' heart rhythms synchronize when a choir sings.[8]

In 1890 William James, often called the father of American psychology, proposed that emotions were names we give to physiological sensations in our bodies.[9] When your heart starts pounding, this physical sensation gives rise to the emotion of fear. You do not get scared first, followed by your heart beating harder and faster. When your heart pounds, you become scared. Researchers recently have found support for James's theory. Using functional neuroimaging techniques, researchers can see that the part of the brain that processes internal sensations including the heartbeat (the anterior insula) is also important in processing emotions.

Interoception is the ability to feel your heartbeat (along with other internal sensations). In contrast, exteroception are signals you receive from the outside world, for example, through sight or sound. Studies are finding that people with enhanced interoceptive accuracy—those more able to feel their heart beating—experience emotions with greater intensity. Individuals with enhanced interoception have greater activation of the anterior insula in the brain. Training people to better detect their own heartbeat, improving their interoceptive accuracy, is now being studied as a way to reduce anxiety and panic attacks.[10]

New research into this heart-brain connection may be the beginning of a scientific shift that is more aligned with the beliefs of our ancient ancestors and with modern cultural views of the heart. The heart is no longer being viewed merely as a pump. The heart influences our emotional vitality, and it shares the role with the brain of ensuring our mental, spiritual, and physical health. It turns out that the heart is a significant player in the way we experience emotions and make decisions.

Chapter Thirty-Six

THE FUTURE HEART

It's the intersection of technology and liberal arts that makes our hearts sing.

Steve Jobs

WHAT'S IN STORE for heart disease prevention and repairing broken hearts in the twenty-first century? The age of "personalized medicine" is upon us. It is now realistic and affordable to screen one's entire genome—your complete set of genes—for susceptibility to heart disease, specific cancers, infectious diseases, and more.[1] Genomic medicine considers the uniqueness of your individual DNA configuration. Differences in genetic makeup can determine what diseases you may be at risk for in the future as well as the medicine to which your disease condition will be most responsive. Genomic screening will identify people at risk for heart disease, and they can receive primary prevention therapies long before heart attack or heart failure may happen.

These advances in genetic identification of future risk for heart disease will lead to gene-informed personalized prevention, and gene-informed therapy, or smart therapy. As an example, surgeons at the University of Michigan Medical Center inserted genetically modified cells into the liver of a twenty-nine-year-old woman who had a genetic defect preventing her liver from removing low-density lipoprotein (LDL) cholesterol from the blood—the cholesterol particle that migrates into the coronary artery wall causing atherosclerotic plaque development.[2] She had suffered a heart attack at sixteen years of age. With the new cells, her liver is able to better remove LDL cholesterol from the blood, potentially reducing her risk of future heart attacks.

■ ■ ■

The worst time to have a heart attack is during a game of charades.

Demetri Martin

Can a damaged heart be repaired like new? Metaphorically we speak of mending a broken heart, but after a heart attack, a patient is left with scar tissue where heart muscle cells have died. Unlike salamanders, humans cannot regenerate heart muscle. However, researchers from King's College in London recently demonstrated that genetic therapy can induce human heart cells to regenerate after a heart attack. The researchers injected a small piece of genetic material (human microRNA-199) into the hearts of pigs (their hearts are a lot like human hearts) after experimentally induced heart attacks.[3] One month later there were significant improvements in heart muscle mass and function. In the near future, similar genetic therapies may induce heart muscle regeneration in humans who suffer a heart attack or other damage to the heart, such as from chemotherapy or infection.

Physician-scientists have recently developed a procedure in which a patient's stem cells can be harvested and injected into scar tissue after a heart attack to transform this tissue into living heart muscle.[4] Stem cells are cells in the adult body that can be induced to change into many different cell types to repair the body, including brain and muscle cells. Recent trials have tested the procedure on heart attack survivors, showing reductions in infarct size due to regeneration of new heart muscle within three months after a heart attack. The procedure involves harvesting stem cells from a patient's bone marrow, multiplying them artificially in the laboratory, and injecting them into the site of the heart injury to repair the damage. In the near future, a mechanical ventricular assist device may provide a bridge to support patients with severe acute heart failure and help their damaged heart survive while injected stem cell therapy replaces their lost heart muscle.

Three-dimensional printing technology is now being used to grow heart tissue by seeding a mix of human cells (heart muscle cells, smooth muscle cells, and endothelial cells; all from human stem cells) onto a 1-micron-resolution scaffold (for scale perspective, a human hair is 70 microns in diameter).[5] The cells organize on the scaffold to create synchronously beating heart tissue. Researchers have placed these cells onto a mouse heart that recently suffered a heart attack, and the laboratory-grown muscle improved heart function. Because the heart cannot create new muscle cells after a heart attack, this technique could be a breakthrough for reducing heart failure after a heart event.

Better yet, why not use the already present scaffolding of a whole heart (such as a pig heart) to grow a whole new heart on?[6] Researchers are studying chemical decellularizing of a whole heart (human or pig), preserving its 3D architecture and vascularity; referred to as a structurally intact decellularized extracellular matrix (dCEM).[7] The theory is that vessels and valves of the heart are maintained, and the patient's heart muscle cells are grown onto the skeleton or scaffolding. Someday this may lead to growing whole personalized hearts for patients with broken ones.

Advances in 3D printing technology will allow physician-scientists to create heart valves that are an exact fit that is customized to each patient. If a patient's native valve is damaged, significantly leaking, or too stiff, an exact replacement can be made and implanted.

■ ■ ■

Instead of rebuilding a damaged heart muscle, promising research is focusing on preventing a heart attack. This is primary prevention as opposed to secondary prevention, which is preventing someone who has already had a heart attack from having another one. What if we could give a primary prevention "vaccine" to those at high risk for future heart disease? One example just approved for use is inclisiran, a medicine that produces long-acting RNA interference (think gene shut down) in liver cells that make cholesterol.[8] This compound can be given twice a year to people with familial hypercholesterolemia (genetically high cholesterol) who develop heart disease as early as their teens. Even better, recent studies in primates are showing that a single injection of CRISPR DNA base editors can produce lifelong reduced liver cholesterol production—a one-and-done genome editing medicine for hypercholesterolemia and heart disease.[9] CRISPR stands for clustered regularly interspaced short palindromic repeats, which can be programmed to target specific stretches of genetic code and to edit DNA at precise locations.

Nanobots (cell-sized robots) are being developed for targeted medical therapies.[10] An example of their futuristic application in cardiology is the development of catheter devices that can use nano-bubbles to push nanobots through a blood clot in a coronary artery, allowing faster penetration of drugs that break down the clot, thereby minimizing the damage of a heart attack.

Biological pacemakers may be a near future alternative to the current implanted electronic devices.[11] Biological pacemakers are cells or genes implanted or injected into the heart that produce electrical stimuli, mimicking the heart's natural pacemaker cells. When the heart's main pacemaker, the sinoatrial node, stops working, it can lead to slow heart rates that are insufficient to support circulation. An electronic pacemaker can be surgically placed into the patient to speed up the heart rate and improve circulation. As a biological alternative, techniques for transferring genes that change working myocardial cells into a surrogate sinoatrial node are being developed. A pacemaker could be made out of one's own heart muscle cells. Even as electronic pacemakers become smaller and more advanced, biological pacemakers might expand the therapeutic armamentarium for a misfiring heart.

Cardiac xenotransplantation may yet become a reality.[12] In 2016, researchers at the U.S. National Institutes of Health reported keeping a genetically engineered pig's heart beating in a baboon for three years. This is a headline-grabbing story, but there are serious implications for this research. Every year several million patients die worldwide because of the lack of human donor hearts available for transplant. Scientists are working on an alternative: hearts from other animals. Although some might condemn this idea as unnatural, remember that the alternative is death. When human-to-human heart transplants were first performed, the morality of the procedure was vigorously questioned, but now it has become routine, with more than eight thousand human heart transplants each year worldwide. The question someday may be whether you will love your significant other more if you receive a loyal dog heart instead of an obstinate pig heart. David Bennet, a fifty-seven-year-old man not considered a candidate for a human heart transplant, became the first successful human recipient of a pig heart. The pig heart had ten genes modified to avoid an aggressive immune response from his body, and Bennet survived two months. Research into gene-edited pig hearts as another source for a viable transplant alternative is ongoing.

Robotic cardiac surgery, also called closed-chest heart surgery, is performed by inserting small robot-controlled surgical instruments through very small incisions in the chest. Current open-heart surgery techniques require surgeons to "crack the chest"—literally split the sternum—and patients have a "zipper" chest scar.[13] Increasing use of robotic techniques enable surgeons to perform less invasive heart surgeries. These procedures are sometimes called da Vinci surgeries because that is the name of the manufacturer of the robots usually used for these procedures. One wonders what Leonardo would think. Performing repairs of heart valves, repairing holes in the heart, and removing heart tumors are examples of da Vinci surgeries that result in improved outcomes, quicker recovery, and shorter hospital stays.

Heart disease prevalence and costs of treatments are projected to increase substantially over the next twenty years because of an obesity epidemic in the United States and increased life spans due to medical advancements. This is inevitable unless people change their unhealthy lifestyles. Future directions in cardiovascular research will include identifying those at risk for heart disease earlier, developing treatments to prevent future heart events, repairing or replacing broken hearts, and exploring the heart-brain connection to better safeguard our hearts physically and emotionally.

AFTERWORD

To give pleasure to a single heart by a single act is better than
a thousand heads bowing in prayer.
Mahatma Gandhi

A good head and a good heart are always a formidable
combination.
Nelson Mandela

A heart is not judged by how much you love; but by how much
you are loved by others.
Wizard of Oz

WHAT HOLDS our life forces? How do we love? Where is our
spirituality that nurtures our capacity for good and evil? These
questions have fascinated humans for twenty thousand years.
In this book, I have explored the continuity of curiosity about
the heart across time and civilizations, philosophically, artisti-
cally, and scientifically. The heart occupies a unique place in the
cultural and religious history of humanity. The heart has been
the beating force at the center of basic human emotions: love
and passion, pain and suffering. The heart was where the soul
and conscience resided, and even our reasoning was presumed
to be a function of the heart.

Since humans first recorded their thoughts, most civilizations believed the heart was the most important organ in the human body. Societies elevated the heart to the position held today by the brain: the ruler of the body and the source of its power. For thousands of years, it was believed that only through the heart could one connect with God. Symbolically, the heart came to represent love, piety, fidelity, courage, friendship, and romance.

Today we believe that the brain controls our body, including the function of the heart. The heart is the first to react to signals from the brain, but the brain is the first to receive the effects of the circulation of blood from the heart. If that weren't the case, we might pass out when we stand quickly. The emotions we feel in our brain reverberate in our heart, and the resulting physical sensations are manifestations of the heart's response. It is this interdependence that has led humans to argue for thousands of years about where the soul is located in the body. When we think of the brain, we envision a mass of cold gray matter, not a warm beating organ that signifies we're alive.

Science and medicine have shaped our present-day conception of the heart. William Harvey's discovery that the heart was nothing more than a pump to circulate the blood pushed thinkers to relegate the heart to the status of just another organ in our body necessary for survival. It was no longer the center of our being but just a muscle that pumped blood containing oxygen to the cells of our body. However, when a patient dies and I must pronounce the legal time of death, it is not when the person is brain dead but is the moment when the person's heart has stopped. When an obstetrician first lets a woman hear the heartbeat of the child in her womb, it affirms the beginning of a new life.

Transplanting a heart into another human being has become commonplace. That we can morally take this beating organ from one human body and place it in another shows how disconnected we have become with the heart as nothing more than a pump. Can you imagine what ancient Egyptians or medieval Christians would have thought?

New research into the heart-brain connection may be the beginning of a scientific shift. More in line with the historic beliefs of our ancient ancestors and modern cultural views, the heart is no longer viewed merely as a pump but is again recognized as part of the emotional vitality that ensures our mental, spiritual, and physical health. Research suggests that the pattern and stability of the heart's rhythm influences higher brain centers and affects psychological factors such as attention, motivation, pain perception, and emotional processing.

The heart continues to play a central role in our cultural iconography. The heart remains a timeless metaphor for the most precious thing we humans have—love. It remains one of the most iconic and widespread symbols in our daily lives. The heart symbol signifies our happiness and health. It is with our heart that we know and feel. In modern times, we act as if we have two hearts: the physiological heart that keeps us alive and the symbolic heart that defines our emotions, desires, courage, and connection to one another. Our heart remains our center.

My heart-felt thanks for taking the time to read this curious history of the heart. In closing, I leave you with a favorite quote from French philosopher and mathematician Blaise Pascal (from Pensées, 1658): "The heart has its reasons, which reason knows not of." I believe Pascal was saying that we know certain things to be true, and that we come to know these truths not through logical reasoning but because we come to believe them with all our heart.

ACKNOWLEDGMENTS

MANY PEOPLE helped make this book a reality. My author friend Tom Barbash convinced me I could, and should, write this book. My first editor, Rachel Lehmann-Haupt of StoryMade Studio, made me tear apart what I thought was a finished book and rebuild it into something so much better. Special thanks to Doctors Bill Harris, Nick Langan, Paul Mather, Roi Altit, and Thad Waites for their advice with early drafts of this manuscript and for assisting me with research. My dear wife Ann acted as my ever-present reader, editor, adviser, and cheerleader. The wonderful people at Columbia University Press helped me place this book in your hand. And finally, my heart-felt thanks to my patients over the years. You were my motivation for loving this subject and wanting to share the heart's curious history.

NOTES

INTRODUCTION

1. William Harvey, *Exercitationes de Generatione Animalium (On Animal Generation)*, 1651, from Exercise 52.
2. Rollin McCraty, Mike Atkinson, Dana Tomasino, and Raymond Trevor Bradley, "The Coherent Heart: Heart–Brain Interactions, Psychophysiological Coherence, and the Emergence of System-Wide Order," *Integral Review* 5, no. 2 (December 2009): 10–115.
3. Ross Toro, *Leading Causes of Death in the US: 1900–Present* (Infographic), July 1, 2012, https://www.livescience.com/21213-leading-causes-of-death-in-the-u-s-since-1900-infographic.html.
4. Irene Fernández-Ruiz, "Breakthrough in Heart Xenotransplantation," *Nature Reviews Cardiology* 16, no. 2 (February 2019): 69.
5. Moo-Sik Lee, Andreas J. Flammer, Lilach O. Lerman, and Amir Lerman, "Personalized Medicine in Cardiovascular Diseases," *Korean Circulation Journal* 42, no. 9 (September 2012): 583–91.

1. THE HEART MEANS LIFE

1. N. K. Sanders, *The Epic of Gilgamesh* (London: Penguin, 1972).
2. Stephanie Dalley, *Myths from Mesopotamia: Creation, the Flood, Gilgamesh, and Others* (Oxford: Oxford University Press, 1989).

3. Spell 30, Book of the Dead, Papyrus of Ani, 1240 BCE. In Raymond Oliver Faulkner, *The Ancient Egyptian Book of the Dead* (London: British Museum Press, 2010).

4. John F. Nunn, *Ancient Egyptian Medicine* (London: British Museum Press, 1996).

5. Kaoru Sakatani, "Concept of Mind and Brain in Traditional Chinese Medicine," *Data Science Journal* 6 (Suppl., 2007): S220–24.

6. Guan Zhong, *Guanzi*, chapter 36, "Techniques of the Heart," in Xiang Liu and W Allyn Rickett, *Guanzi: Political, Economic, and Philosophical Essays from Early China* (Princeton, NJ: Princeton University Press. 1985).

7. *Huainanzi* IX and XX, in Liu An, King of Huinan, *The Huainanzi: A Guide to the Theory and Practice of Government in Early Han China*, ed. and trans. John S. Major, Sarah A. Queen, Andrew Seth Meyer, and Harold D. Roth (New York: Columbia University Press, 2010).

8. Li Yuheng, *Unfolding the Mat with Enlightening Words* (Tuipeng Wuyu), Ming Dynasty, 1570, https://classicalchinesemedicine.org/heart-selected-readings.

9. Li Ting, *A Primer of Medicine* (Yixue Rumen), 1575, https://classicalchinese medicine.org/heart-selected-readings.

10. K. Chimin Wong and Wu Lien-Teh, *History of Chinese Medicine: Being a Chronicle of Medical Happenings in China from Ancient Times to the Present Period*, 2nd ed. (Shanghai, China: National Quarantine Service, 1936 and reprinted by Taipei, Taiwan: Southern Materials Center), 35.

11. Kishor Patwardhan, "The History of the Discovery of Blood Circulation: Unrecognized Contributions of Ayurveda Masters," *Advances in Physiology Education* 36, no. 2 (2012): 77–82.

2. HEART AND SOUL

1. Amentet Neferet, *Ancient Egyptian Dictionary*, accessed December 2021, https://seshkemet.weebly.com/dictionary.html.

2. Kaoru Sakatani, "Concept of Mind and Brain in Traditional Chinese Medicine," *Data Science Journal* 6 (Suppl., 2007): S220–24.

3. V. Jayaram, "The Meaning and Significance of Heart in Hinduism," 2019, https://www.hinduwebsite.com/hinduism/essays/the-meaning-and-significance-of-heart-in-hinduism.asp.

4. C. R. S. Harris, *The Heart and Vascular System in Ancient Greek Medicine: From Alcmaeon to Galen* (Oxford: Oxford University Press, 1973).
5. Harris, *The Heart and Vascular System in Ancient Greek Medicine.*

3. THE HEART AND GOD

1. Kenneth G. Zysk, *Religious Medicine: The History and Evolution of Indian Medicine* (London: Transaction, 1993).
2. Marjorie O'Rourke Boyle, *Cultural Anatomies of the Heart in Aristotle, Augustine, Aquinas, Calvin, and Harvey* (London: Palgrave Macmillan, 2018).

4. AN EMOTIONAL HEART

1. Kenneth G. Zysk, *Religious Medicine: The History and Evolution of Indian Medicine* (London: Transaction, 1993).
2. C. R. S. Harris, *The Heart and Vascular System in Ancient Greek Medicine: From Alcmaeon to Galen* (Oxford: Oxford University Press, 1973).
3. Helen King, *Greek and Roman Medicine* (Bristol: Bristol Classical Press, 2001).

5. ANCIENT UNDERSTANDING OF THE PHYSICAL HEART

1. C. R. S. Harris, *The Heart and Vascular System in Ancient Greek Medicine: From Alcmaeon to Galen* (Oxford: Oxford University Press, 1973).
2. Marjorie O'Rourke Boyle, *Cultural Anatomies of the Heart in Aristotle, Augustine, Aquinas, Calvin, and Harvey* (London: Palgrave Macmillan, 2018).
3. Celsus, *Prooemium: De Medicina*, Book 1, ed. W. G. Spencer (Cambridge, MA: Harvard University Press, 1971).
4. Harris, *The Heart and Vascular System in Ancient Greek Medicine*, 271.
5. Helen King, *Greek and Roman Medicine* (London: Bristol Classical Press, 2001).
6. Harris, *The Heart and Vascular System in Ancient Greek Medicine*, 271.
7. Galen, *On the Affected Parts*, V:1,2.

6. ANCIENT HEART DISEASE

1. Adel H. Allam, Randall C. Thompson, L. Samuel Wann, Michael I. Miyamoto, and Gregory S. Thomas, "Computed Tomographic Assessment of Atherosclerosis in Ancient Egyptian Mummies," *JAMA* 302, no. 19 (November 2009): 2091–94.

2. Randall C. Thompson, Adel H. Allam, Guido P. Lombardi, L. Samuel Wann, M. Linda Sutherland, James D. Sutherland, Muhammad Al-Tohamy Soliman, Bruno Frohlich, David T. Mininberg, Janet M. Monge, Clide M. Vallodolid, Samantha L. Cox, Gomaa Abd el-Maksoud, Ibrahim Badr, Michael I. Miyamoto, Abd el-Halim Nur el-din, Jagat Narula, Caleb E. Finch, and Gregory S. Thomas, "Atherosclerosis Across 4000 Years of Human History: The Horus Study of Four Ancient Populations," *Lancet* 381, no. 9873 (2013): 1211–22.

3. Andreas Keller, Angela Graefen, Markus Ball, Mark Matzas, Valesca Boisguerin, Frank Maixner, Petra Leidinger, Christina Backes, Rabab Khairat, Michael Forster, Björn Stade, Andre Franke, Jens Mayer, Jessica Spangler, Stephen McLaughlin, Minita Shah, Clarence Lee, Timothy T. Harkins, Alexander Sartori, Andres Moreno-Estrada, Brenna Henn, Martin Sikora, Ornella Semino, Jacques Chiaroni, Siiri Roostsi, Natalie M. Myres, Vicente M. Cabrera, Peter A. Underhill, Carlos D. Bustamante, Eduard Egarter Vigl, Marco Samadelli, Giovanna Cipollini, Jan Haas, Hugo Katus, Brian D. O'Connor, Marc R. J. Carlson, Benjamin Meder, Nikolaus Blin, Eckart Meese, Carsten M. Pusch, and Albert Zink, "New Insights Into the Tyrolean Iceman's Origin and Phenotype as Inferred by Whole-Genome Sequencing," *Nature Communications* 3 (February 2012): 698.

7. THE DARK AGES

1. Heather Webb, *The Medieval Heart* (New Haven, CT: Yale University Press, 2010).

2. Piero Camporesi, *The Incorruptible Flesh: Bodily Mutation and Mortification in Religion and Folklore*, trans. Tania Croft-Murray (New York: Cambridge University Press, 1988).

3. Camporesi, *The Incorruptible Flesh*, 5.

4. Bertrand Mafart, "Post-Mortem Ablation of the Heart: A Medieval Funerary Practice. A Case Observed at the Cemetery of Ganagobie

Priory in the French Department of Alpes De Haute Provence," *International Journal of Osteoarchaeology* 14, no. 1 (2004): 67–73.

5. Katie Barclay, "Dervorgilla of Galloway (abt 1214–abt 1288)," *Women's History Network*, August 15, 2010, https://womenshistorynetwork.org /dervorgilla-of-galloway-abt-1214-abt-1288/.

6. Marjorie O'Rourke Boyle, "Aquinas's Natural Heart," *Early Science and Medicine* 18, no. 3 (2013): 266–90.

8. THE ISLAMIC GOLDEN AGE

1. Hawa Edriss, Brittany N. Rosales, Connie Nugent, Christian Conrad, and Kenneth Nugent, "Islamic Medicine in the Middle Ages," *American Journal of the Medical Sciences* 354, no. 3 (September 2017): 223–29.

2. André Silva Ranhel and Evandro Tinoco Mesquita, "The Middle Ages Contributions to Cardiovascular Medicine," *Brazilian Journal of Cardiovascular Surgery* 31, no. 2 (April 2016): 163–70.

3. Rachel Hajar, "Al-Razi: Physician for All Seasons," *Heart Views* 6, no. 1 (2005): 39–43.

4. Hajar, "Al-Razi: Physician for All Seasons," 41.

9. THE VIKING COLD HJARTA

1. Snorre Sturlason, *Heimskringla—The Norse King Sagas* (Redditch, UK: Read Books, 2011).

10. AMERICAN HEART SACRIFICE

1. Michael D. Coe and Rex Koontz, *Mexico: From the Olmecs to the Aztecs* (London: Thames and Hudson, 2008).

2. James Maffie, "Aztec Philosophy," *Internet Encyclopedia of Philosophy*, April 3, 2022, https://iep.utm.edu/aztec-philosophy/.

3. Molly H. Bassett, *The Fate of Earthly Things: Aztec Gods and God-Bodies* (Austin: University of Texas Press, 1980).

4. Gabriel Prieto, John W. Verano, Nicolas Goepfert, Douglas Kennett, Jeffrey Quilter, Steven LeBlanc, Lars Fehren-Schmitz, Jannine Forst, Mellisa Lund, Brittany Dement, Elise Dufour, Olivier Tombret, Melina Calmon, Davette Gadison, and Khrystyne Tschinkel, "A Mass

Sacrifice of Children and Camelids at the Huanchaquito-Las Llamas Site, Moche Valley, Peru," *PLoS One* 14, no. 3 (2019): e0211691.

5. Bernal Diaz Del Castillo, *The True History of the Conquest of New Spain* (London: Penguin Classics, 2003), 104.

6. Haverford College, Intro to Environmental Anthropology Class, *The Gwich'in People: Caribou Protectors*, December 2021, https://anthro281 .netlify.app.

11. THE HEART RENAISSANCE

1. William W. E. Slights, "The Narrative Heart of the Renaissance," *Renaissance and Reformation* 26, no. 1 (2002): 5–23.

2. Marco Cambiaghi and Heidi Hausse, "Leonardo da Vinci and His Study of the Heart," *European Heart Journal* 40, no. 23 (2019): 1823–26.

3. Mark E. Silverman, "Andreas Vesalius and *de Humani Corporis Fabrica*," *Clinical Cardiology* 14 (1991): 276–79.

12. HITHER AND THITHER

1. Thomas Fuchs, *Mechanization of the Heart: Harvey and Descartes* (Rochester, NY: University of Rochester Press, 2001).

2. William Harvey, *Exercitatio Anatomica de Motu Cordis et Sanguinis in Animalibus*, chap. 13.

3. William Harvey, *Lectures on the Whole of Anatomy*, 92.

4. William Harvey, *Exercitationes de Generatione Animalium (On Animal Generation)* (1651), Exercise 52.

5. W. Bruce Fye, "Profiles in Cardiology: René Descartes," *Clinical Cardiology* 26, no. 1 (2003): 49–51.

6. Descartes, *Traité de l'homme* (Treatise on Man), 1664.

13. THE HEART IN ART

1. Pierre Vinken, "How the Heart Was Held in Medieval Art," *Lancet* 358, no. 9299 (2001): 2155–57.

2. Adi Kalin, "Frau Minne hat sich gut gehalten," *NZZ*, November 25, 2009, https://www.nzz.ch/frau_minne_hat_sich_gut_gehalten-ld.930946?.

3. Gordon Bendersky, "The Olmec Heart Effigy: Earliest Image of the Human Heart," *Perspectives in Biology and Medicine* 40, no. 3 (Spring 1997): 348–61.

14. THE HEART IN LITERATURE

1. William W. E. Slights, "The Narrative Heart of the Renaissance," *Renaissance and Reformation* 26, no. 1 (2002): 5–23.

15. THE HEART IN MUSIC

1. Coding in Tune, "Most Used Words in Lyrics by Genre," April 2018, https://codingintune.com/2018/04/09/statistics-most-used-words-in-lyrics-by-genre/.

16. HEART RITUALS

1. Ambrosius Aurelius Theodosius Macrobius, *Seven Books of the Saturnalia*, accessed April 2022, https://www.loc.gov/item/2021667911/.
2. T. Christian Miller, "A History of the Purple Heart," *NPR*, September 2010, https://www.npr.org/templates/story/story.php?storyId=129711544.

18. HEART ANATOMY

1. Xiaoya Ma, Peiyun Cong, Xianguang Hou, Gregory D. Edgecombe, and Nicholas J. Strausfeld, "An Exceptionally Preserved Arthropod Cardio-vascular System from the Early Cambrian," *Nature Communications* 5 (2014): 3560.
2. Brandon Specktor, "Evolution Turned This Fish Into a 'Penis with a Heart.' Here's How," *Live Science*, August 3, 2020, https://www.livescience.com/anglerfish-fusion-sex-immune-system.html.
3. Jeremy B. Swann, Stephen J. Holland, Malte Petersen, Theodore W. Pietsch, and Thomas Boehm, "The Immunogenetics of Sexual Parasit-ism," *Science* 369, no. 6511 (2020): 1608–15.

21. THE HEART'S ELECTRICAL SYSTEM

1. W. Bruce Fye, "A History of the Origin, Evolution, and Impact of Electrocardiography," *American Journal of Cardiology* 73, no. 13 (1994): 937–49.

22. WHAT IS AN EKG?

1. O. Aquilina, "A Brief History of Cardiac Pacing," *Images in Paediatriic Cardiology* 8, no. 2 (April–June 2006):17–81.

23. WHAT IS BLOOD PRESSURE?

1. Nature Editors, "Samuel Siegfried Karl von Basch (1837–1905)," *Nature* 140 (1937): 393–94.
2. World Health Organization, "Hypertension," August 25, 2021, https://www.who.int/news-room/fact-sheets/detail/hypertension.
3. Timothy Bishop and Vincent M. Figueredo, "Hypertensive Therapy: Attacking the Renin-Angiotensin System," *Western Journal of Medicine* 175, no. 2 (August 2001): 119–24.
4. William Osler, "An Address on High Blood Pressure: Its Associations, Advantages, and Disadvantages: Delivered at the Glasgow Southern Medical Society," *British Medical Journal* 2, no. 2705 (November 2, 1912): 1173–77.
5. G. Antonakoudis, L. Poulimenos, K. Kifnidis, C. Zouras, and H. Antonakoudis, "Blood Pressure Control and Cardiovascular Risk Reduction," *Hippokratia* 11, no. 3 (July 2007): 114–19.

24. WHAT IS HEART FAILURE?

1. A. Perciaccante, M. A. Riva, A. Coralli, P. Charlier, and R. Bianucci, "The Death of Balzac (1799–1850) and the Treatment of Heart Failure During the Nineteenth Century," *Journal of Cardiac Failure* 22, no. 11 (2016): 930–33.
2. Raffaella Bianucci, Robert D. Loynes, M. Linda Sutherland, Rudy Lallo, Gemma L. Kay, Philippe Froesch, Mark J. Pallen, Philippe Charlier, and Andreas G. Nerlich, "Forensic Analysis Reveals Acute Decompensation

of Chronic Heart Failure in a 3500-Year-Old Egyptian Dignitary," *Journal of Forensic Sciences* 61, no. 5 (September 2016): 1378–81.

3. Roberto Ferrari, Cristina Balla, and Alessandro Fucili, "Heart Failure: An Historical Perspective," *European Heart Journal Supplements* 18 (Suppl. G, 2016): G3–G10.

25. WHAT IS "HAVING A CORONARY"?

1. W. F. Enos, R. H. Holmes, and J. Beyer, "Coronary Disease Among United States Soldiers Killed in Action in Korea: Preliminary Report," *JAMA* 152, no. 12 (1953):1090–93.

2. J. J. McNamara, M. A. Molot, J. F. Stremple, and R. T. Cutting, "Coronary Artery Disease in Combat Casualties in Vietnam," *JAMA* 216, no. 7 (1971):1185–87.

3. Manoel E. S. Modelli, Aurea S. Cherulli, Lenora Gandolfi, and Riccardo Pratesi, "Atherosclerosis in Young Brazilians Suffering Violent Deaths: A Pathological Study," *BMC Research Notes* 4 (2011): 531.

4. James B. Herrick, "Clinical Features of Sudden Obstruction of the Coronary Arteries," *JAMA* 59 (1912): 2015–20.

26. SEX, RACE, AND ETHNICITY IN HEART DISEASE

1. U.S. Department of Health and Human Services Office of Minority Health, "Heart Disease and African Americans," January 31, 2022, https://minorityhealth.hhs.gov/omh/browse.aspx?lvl=4&lvlid=19.

2. Centers for Disease Control, "Disparities in Premature Deaths from Heart Disease," February 19, 2004, https://www.cdc.gov/mmwr/preview/mmwrhtml/mm5306a2.htm.

3. World Health Organization, "The Top 10 Causes of Death," December 2020, https://www.who.int/news-room/fact-sheets/detail/the-top-10-causes-of-death.

4. World Health Organization, "Cardiovascular Disease," June 2021, https://www.who.int/news-room/fact-sheets/detail/cardiovascular-diseases-(cvds).

5. Centers for Disease Control and Prevention, "Preventing 1 Million Heart Attacks and Strokes," September 6, 2018, https://www.cdc.gov/vitalsigns/million-hearts/.

6. American Heart Association, "Championing Health Equity for All," April 2022, https://www.heart.org/en/about-us/2024-health-equity-impact -goal.

7. American Heart Association, "Championing Health Equity for All"; American College of Cardiology, "Cover Story | Health Disparities and Social Determinants of Health: Time for Action," June 11, 2020, https:// bluetoad.com/publication/?m=14537&i=664103&p=1&ver=html5.

8. A. H. E. M. Maas and Y. E. A. Appelman, "Gender Differences in Coronary Heart Disease," *Netherlands Heart Journal* 18, no. 12 (December 2010): 598–602.

9. Alan S. Go, Dariush Mozaffarian, Véronique L. Roger, Emelia J. Benjamin, Jarett D. Berry, William B. Borden, Dawn M. Bravata, Shifan Dai, Earl S. Ford, Caroline S. Fox, Sheila Franco, Heather J. Fullerton, Cathleen Gillespie, Susan M. Hailpern, John A. Heit, Virginia J. Howard, Mark D. Huffman, Brett M. Kissela, Steven J. Kittner, Daniel T. Lackland, Judith H. Lichtman, Lynda D. Lisabeth, David Magid, Gregory M. Marcus, Ariane Marelli, David B. Matchar, Darren K. McGuire, Emile R. Mohler, Claudia S. Moy, Michael E. Mussolino, Graham Nichol, Nina P. Paynter, Pamela J. Schreiner, Paul D. Sorlie, Joel Stein, Tanya N. Turan, Salim S. Virani, Nathan D. Wong, Daniel Woo, and Melanie B. Turner, "Heart Disease and Stroke Statistics—2013 Update: A Report from the American Heart Association," *Circulation* 127, no. 1 (January 2013): e6–e245.

10. American Heart Association, "The Facts About Women and Heart Disease," April 2022, https://www.goredforwomen.org/en/about-heart -disease-in-women/facts.

27. SUDDEN DEATH OF AN ATHLETE

1. Michael S. Emery and Richard J. Kovacs, "Sudden Cardiac Death in Athletes," *JACC Heart Failure* 6, no. 1 (2018): 30–40.

2. Meagan M. Wasfy, Adolph M. Hutter, and Rory B. Weiner, "Sudden Cardiac Death in Athletes," *Methodist Debakey Cardiovascular Journal* 12, no. 2 (2016): 76–80.

3. American Heart Association, "Recommendations for Physical Activity in Adults and Kids," last reviewed April 18, 2018, https://www.gored forwomen.org/en/healthy-living/fitness/fitness-basics/aha-recs-for -physical-activity-in-adults.

29. ENLIGHTENMENT AND THE AGE OF REVOLUTION

1. Luis-Alfonso Arráez-Aybar, Pedro Navia-Álvarez, Talia Fuentes-Redondo, and José-L Bueno-López, "Thomas Willis, a Pioneer in Translational Research in Anatomy (on the 350th Anniversary of Cerebri Anatome)," *Journal of Anatomy* 226, no. 3 (March 2015): 289–300.

2. John B. West, "Marcello Malpighi and the Discovery of the Pulmonary Capillaries and Alveoli," *American Journal of Physiology, Lung Cellular and Molecular Physiology* 304, no. 6 (2013): L383–90.

3. Edmund King, "Arthur Coga's Blood Transfusion (1667)," *Public Domain Review*, April 15, 2014, https://publicdomainreview.org/collection/arthur -coga-s-blood-transfusion-1667.

4. Marios Loukas, Pamela Clarke, R. Shane Tubbs, and Theodoros Kapos, "Raymond de Vieussens," *Anatomical Science International* 82, no. 4 (2007): 233–36.

5. Max Roser, Esteban Ortiz-Ospina, and Hannah Ritchie, "Life Expectancy," *Our World in Data*, last revised October 2019, https://ourworldindata .org/life-expectancy.

6. Maria Rosa Montinari and Sergio Minelli, "The First 200 Years of Cardiac Auscultation and Future Perspectives," *Journal of Multidisciplinary Healthcare* 12 (2109): 183–89.

7. Ariel Roguin, "Rene Theophile Hyacinthe Laënnec (1781–1826): The Man Behind the Stethoscope," *Clinical Medicine and Research* 4, no. 3 (2006): 230–35.

8. William Heberden, "Some Account of a Disorder of the Breast," *Medical Transactions. The Royal College of London* 2 (1772): 59–67.

9. Although the name "catgut" implies the use of guts from cats, the word is derived from *kitgut*, the string used on a fiddle or "kit." The first known absorbable catgut sutures were made from intestines of sheep or cows. They were being used as medical sutures as early as the third century by Galen in Rome. Today catgut has largely been replaced by absorbable synthetic polymers.

10. L. Rehn, "Ueber penetrierende Herzwunden und Herznaht," *Arch Klin Chir* 55, no. 315 (1897): 315–29.

11. Paul, "Door 23: The Heart of a King," *Geological Society of London* (blog), December 23, 2014, https://blog.geolsoc.org.uk/2014/12/23/the-heart-of -a-king/.

12. Stacey Conradt, "Mary Shelley's Favorite Keepsake: Her Dead Husband's Heart," *Mental Floss*, July 8, 2015, https://www.mentalfloss.com/article /65624/mary-shelleys-favorite-keepsake-her-dead-husbands-heart.

30. THE TWENTIETH CENTURY AND HEART DISEASE

1. Ross Toro, *Leading Causes of Death in the US: 1900–Present* (Infographic), July 1, 2012, https://www.livescience.com/21213-leading-causes-of-death -in-the-u-s-since-1900-infographic.html.
2. World Health Organization, "Cardiovascular Diseases," June 2021, https:// www.who.int/news-room/fact-sheets/detail/cardiovascular-diseases-(cvds).
3. Toro, "Leading Causes of Death in the US."
4. Rachel Hajar, "Coronary Heart Disease: From Mummies to 21st Century," *Heart Views* 18, no. 2 (2017): 68–74.
5. W. P. Obrastzow and N. D. Staschesko, "Zur Kenntnissder Thrombose der Coronararterien des Herzens," *Zeitschrift für klinische Medizin* 71 (1910): 12.

31. ASPIRIN

1. Dawn Connelly, "A History of Aspirin," *Pharmaceutical Journal*, September 2014, https://pharmaceutical-journal.com/article/infographics /a-history-of-aspirin.
2. Jonathan Miner and Adam Hoffhines, "The Discovery of Aspirin's Anti-thrombotic Effects," *Texas Heart Journal* 34, no. 2 (2007): 179–86.

32. THE TWENTIETH CENTURY AND HEART SURGERY

1. Lawrence H. Cohn, "Fifty Years of Open-Heart Surgery," *Circulation* 107, no. 17 (2003): 2168–70; C. W. Lillehei, "The Society Lecture. European Society for Cardiovascular Surgery Meeting, Montpellier, France, September 1992. The Birth of Open-Heart Surgery: Then the Golden Years," *Cardiovascular Surgery* 2, no. 3 (1994): 308–17.
2. Global Observatory on Donation and Transplantation, "Total Heart," April 3, 2022, http://www.transplant-observatory.org/data-charts-and -tables/chart/.

33. THE HEART NOW

1. Centers for Disease Control and Prevention, "Heart Disease Facts," February 7, 2022, https://www.cdc.gov/heartdisease/facts.htm.

34. BROKEN HEART SYNDROME

1. A. Tofield, "Hikaru Sato and Takotsubo Cardiomyopathy," *European Heart Journal* 37, no. 37 (October 2016): 2812.
2. Rienzi Díaz-Navarro, "Takotsubo Syndrome: The Broken-Heart Syndrome," *British Journal of Cardiology* 28 (2021): 30–34.
3. Mahek Shah, Pradhum Ram, Kevin Bryan U. Lo, Natee Sirinvaravong, Brijesh Patel, Byomesh Tripathi, Shantanu Patil, and Vincent M. Figueredo, "Etiologies, Predictors, and Economic Impact of Readmission Within 1 Month Among Patients with Takotsubo Cardiomyopathy," *Clinical Cardiology* 41, no. 7 (July 2018): 916–23.
4. Vincent M. Figueredo, "The Time Has Come for Physicians to Take Notice: The Impact of Psychosocial Stressors on the Heart," *American Journal of Medicine* 122, no. 8 (2009): 704–12.
5. Dean Burnett, "Why Elderly Couples Often Die Together: The Science of Broken Hearts," *Guardian*, January 9, 2015, https://www.theguardian .com/lifeandstyle/shortcuts/2015/jan/09/why-elderly-couples-die -together-science-broken-hearts.

35. THE HEART-BRAIN CONNECTION

1. Vincent M. Figueredo, "The Time Has Come for Physicians to Take Notice: The Impact of Psychosocial Stressors on the Heart," *American Journal of Medicine* 122, no. 8 (2009): 704–12.
2. Annika Rosengren, Steven Hawken, Stephanie Ounpuu, Karen Sliwa, Mohammad Zubaid, Wael A. Almahmeed, Kathleen Ngu Blackett, Chitr Sitthi-amorn, Hiroshi Sato, Salim Yusuf, and INTERHEART investigators, "Association of Psychosocial Risk Factors with Risk of Acute Myocardial Infarction in 11,119 Cases and 13,648 Controls from 52 Countries (the INTERHEART Study): Case-Control Study," *Lancet* 364, no. 9438 (2004): 953–62.
3. Michael Miller, "Emotional Rescue: The Heart-Brain Connection," *Cerebrum* (May 2019): cer-05-19.
4. Rollin McCraty, Mike Atkinson, Dana Tomasino, and Raymond Trevor Bradley, "The Coherent Heart: Heart–Brain Interactions, Psychophysiological Coherence, and the Emergence of System-Wide Order," *Integral Review* 5, no. 2 (December 2009): 10–115; Tara Chand, Meng Li, Hamidreza Jamalabadi, Gerd Wagner, Anton Lord, Sarah Alizadeh,

Lena V. Danyeli, Luisa Herrmann, Martin Walter, and Zumrut D. Sen, "Heart Rate Variability as an Index of Differential Brain Dynamics at Rest and After Acute Stress Induction," *Frontiers in Neuroscience* 14 (July 2020): 645; Sarah Garfinkel, "It's an Intriguing World That Is Opening Up," *The Psychologist* 32 (January 2019): 38–41; Fred Shaffer, Rollin McCraty, and Christopher L. Zerr, "A Healthy Heart Is Not a Metronome: An Integrative Review of the Heart's Anatomy and Heart Rate Variability," *Frontiers in Psychology* 5 (2014): 1040.

5. Ali M. Alshami, "Pain: Is It All in the Brain or the Heart?," *Current Pain and Headache Reports* 23, no. 12 (November 2019): 88.

6. Sirisha Achanta, Jonathan Gorky, Clara Leung, Alison Moss, Shaina Robbins, Leonard Eisenman, Jin Chen, Susan Tappan, Maci Heal, Navid Farahani, Todd Huffman, Steve England, Zixi (Jack) Cheng, Rajanikanth Vadigepalli, and James S. Schwaber, "A Comprehensive Integrated Anatomical and Molecular Atlas of Rat Intrinsic Cardiac Nervous System," *iScience* 23, no. 6 (June 2020): 101140.

7. L. Z. Song, G. E. Schwartz, and L. G. Russek, "Heart-Focused Attention and Heart-Brain Synchronization: Energetic and Physiological Mechanisms," *Alternative Therapies in Health and Medicine* 4, no. 5 (September 1998): 44–52, 54–60, 62.

8. Björn Vickhoff, Helge Malmgren, Rickard Aström, Gunnar Nyberg, Seth-Reino Ekström, Mathias Engwall, Johan Snygg, Michael Nilsson, and Rebecka Jörnsten, "Music Structure Determines Heart Rate Variability of Singers," *Frontiers in Psychology* 4 (July 2013): 334; Apit Hemakom, Katarzyna Powezka, Valentin Goverdovsky, Usman Jaffer, and Danilo P. Mandic, "Quantifying Team Cooperation Through Intrinsic Multi-Scale Measures: Respiratory and Cardiac Synchronization in Choir Singers and Surgical Teams," *Royal Society Open Access* 4, no. 12 (November 2017): 170853.

9. Julian F. Thayer and Richard D. Lane, "Claude Bernard and the Heart-Brain Connection: Further Elaboration of a Model of Neurovisceral Integration," *Neuroscience & Biobehavioral Reviews* 33, no. 2 (2009): 81–88; William James, *The Principles of Psychology* (New York: Henry Holt, 1890).

10. Hugo D. Critchley and Sarah N. Garfinkel, "Interoception and Emotion," *Current Opinion in Psychology* 17 (April 2017): 7–14.

36. THE FUTURE HEART

1. Moo-Sik Lee, Andreas J. Flammer, Lilach O. Lerman, and Amir Lerman, "Personalized Medicine in Cardiovascular Diseases," *Korean Circulation Journal* 42, no. 9 (2012): 583–91; F. Randy Vogenberg, Carol Isaacson Barash, and Michael Pursel, "Personalized Medicine: Part 1: Evolution and Development Into Theranostics," *Pharmacy and Therapeutics* 35, no. 10 (2010): 565–67.

2. M. Grossman, S. E. Raper, K. Kozarsky, E. A. Stein, J. F. Engelhardt, D. Muller, P. J. Lupien, and J. M. Wilson, "Successful Ex Vivo Gene Therapy Directed to Liver in a Patient with Familial Hypercholesterolaemia," *Nature Genetics* 6, no. 4 (1994): 335–41.

3. K. Gabisonia G. Prosdocimo, G. D. Aquaro, L. Carlucci, L. Zentilin, I. Secco, H. Ali, L. Braga, N. Gorgodze, F. Bernini, S. Burchielli, C. Collesi, L. Zandonà, G. Sinagra, M. Piacenti, S. Zacchigna, R. Bussani, F. A. Recchia, and M. Giacca, "MicroRNA Therapy Stimulates Uncontrolled Cardiac Repair After Myocardial Infarction in Pigs," *Nature* 569, no. 7756 (2019): 418–22.

4. Akon Higuchi, Nien-Ju Ku, Yeh-Chia Tseng, Chih-Hsien Pan, Hsing-Fen Li, S. Suresh Kumar, Qing-Dong Ling, Yung Chang, Abdullah A. Alarfaj, Murugan A. Munusamy, Giovanni Benelli, and Kadarkarai Muruga, "Stem Cell Therapies for Myocardial Infarction in Clinical Trials: Bioengineering and Biomaterial Aspects," *Laboratory Investigation* 97 (2017): 1167–79.

5. Shixing Huang, Yang Yang, Qi Yang, Qiang Zhao, and Xiaofeng Ye, "Engineered Circulatory Scaffolds for Building Cardiac Tissue," *Journal of Thoracic Disease* 10 (Suppl. 20; 2018): S2312–28.

6. Brendan Maher, "Tissue Engineering: How to Build a Heart," *Nature* 499 (2013): 20–22.

7. Laura Iop, Eleonora Dal Sasso, Roberta Menabò, Fabio Di Lisa, and Gino Gerosa, "The Rapidly Evolving Concept of Whole Heart Engineering," *Stem Cells International* (2017): 8920940.

8. Frederick J. Raal, David Kallend, Kausik K. Ray, Traci Turner, Wolfgang Koenig, R. Scott Wright, Peter L. J. Wijngaard, Danielle Curcio, Mark J. Jaros, Lawrence A. Leiter, John J. P. Kastelein, and ORION-9 Investigators, "Inclisiran for the Treatment of Heterozygous Familial Hypercholesterolemia," *New England Journal of Medicine* 382, no. 16 (2020): 1520–30.

9. Kiran Musunuru, Alexandra C. Chadwick, Taiji Mizoguchi, Sara P. Garcia, Jamie E. DeNizio, Caroline W. Reiss, Kui Wang, Sowmya Iyer,

Chaitali Dutta, Victoria Clendaniel, Michael Amaonye, Aaron Beach, Kathleen Berth, Souvik Biswas, Maurine C. Braun, Huei-Mei Chen, Thomas V. Colace, John D. Ganey, Soumyashree A. Gangopadhyay, Ryan Garrity, Lisa N. Kasiewicz, Jennifer Lavoie, James A. Madsen, Yuri Matsumoto, Anne Marie Mazzola, Yusuf S. Nasrullah, Joseph Nneji, Huilan Ren, Athul Sanjeev, Madeleine Shay, Mary R. Stahley, Steven H. Y. Fan, Ying K. Tam, Nicole M. Gaudelli, Giuseppe Ciaramella, Leslie E. Stolz, Padma Malyala, Christopher J. Cheng, Kallanthottathil G. Rajeev, Ellen Rohde, Andrew M. Bellinger, and Sekar Kathiresan, "In Vivo CRISPR Base Editing of PCSK9 Durably Lowers Cholesterol in Primates," *Nature* 593, no. 7859 (2021): 429–34.

10. U. Kei Cheang and Min Jun Kim, "Self-Assembly of Robotic Micro- and Nanoswimmers Using Magnetic Nanoparticles," *Journal of Nanoparticle Research* 17 (2015): 145; Jiangfan Yu, Ben Wang, Xingzhou Du, Qianqian Wang, and Li Zhang, "Ultra-Extensible Ribbon-Like Magnetic Microswarm," *Nature Communications* 9, no. 1 (2018): 3260.

11. Eugenio Cingolani, Joshua I. Goldhaber, and Eduardo Marbán, "Next-Generation Pacemakers: From Small Devices to Biological Pacemakers," *Nature Reviews Cardiology* 15, no. 3 (2018): 139–50.

12. Irene Fernández-Ruiz, "Breakthrough in Heart Xenotransplantation," *Nature Reviews Cardiology* 16, no. 2 (February 2019): 69; Martha Längin Tanja Mayr, Bruno Reichart, Sebastian Michel, Stefan Buchholz, Sonja Guethoff, Alexey Dashkevich, Andrea Baehr, Stephanie Egerer, Andreas Bauer, Maks Mihalj, Alessandro Panelli, Lara Issl, Jiawei Ying, Ann Kathrin Fresch, Ines Buttgereit, Maren Mokelke, Julia Radan, Fabian Werner, Isabelle Lutzmann, Stig Steen, Trygve Sjöberg, Audrius Paskevicius, Liao Qiuming, Riccardo Sfriso, Robert Rieben, Maik Dahlhoff, Barbara Kessler, Elisabeth Kemter, Mayulko Kurome, Valeri Zakhartchenko, Katharina Klett, Rabea Kingel, Christian Kupatt, Almuth Falkenau, Simone Reu, Reinhrad Ellgass, Rudolf Herzog, Uli Binder, Günter Wich, Arne Skerra, David Ayares, Alexander Kind, Uwe Schönmann. Franz-Josef Kaup, Christain Hagl, Eckhard Wolf, Nikolai Klymuk, Paolo Brenner, and Jan-Michael Abicht, "Consistent Success in Life-Supporting Porcine Cardiac Xenotransplantation," *Nature* 564, no. 7736 (2018): 430–33.

13. Matteo Pettinari, Emiliano Navarra, Philippe Noirhomme, and Herbert Gutermann, "The State of Robotic Cardiac Surgery in Europe," *Annals of Cardiothoracic Surgery* 6, no. 1 (2017): 1–8.

REFERENCES

Achanta, Sirisha, Jonathan Gorky, Clara Leung, Alison Moss, Shaina Robbins, Leonard Eisenman, Jin Chen, Susan Tappan, Maci Heal, Navid Farahani, Todd Huffman, Steve England, Zixi (Jack) Cheng, Rajanikanth Vadige-palli, and James S Schwaber. "A Comprehensive Integrated Anatomical and Molecular Atlas of Rat Intrinsic Cardiac Nervous System," *iScience* 23, no. 6 (June 2020): 101140. https://doi.org/10.1016/j.isci.2020.101140.

Allam, Adel H., Randall C. Thompson, L. Samuel Wann, Michael I. Miy-amoto, and Gregory S. Thomas. "Computed Tomographic Assessment of Atherosclerosis in Ancient Egyptian Mummies," *JAMA* 302, no. 19 (November 2009): 2091–94.

Alshami, Ali M. "Pain: Is It All in the Brain or the Heart?," *Current Pain and Headache Reports* 23, no. 12 (November 2019): 88.

American College of Cardiology. "Cover Story | Health Disparities and Social Determinants of Health: Time for Action." June 11, 2020. https://www.acc .org/latest-in-cardiology/articles/2020/06/01/12/42/cover-story-health -disparities-and-social-determinants-of-health-time-for-action.

American Heart Association. "Championing Health Equity for All." April 2022. https://www.heart.org/en/about-us/2024-health-equity-impact-goal.

——. "The Facts About Women and Heart Disease." updated April 2022. https://www.goredforwomen.org/en/about-heart-disease-in-women/facts.

——. "Recommendations for Physical Activity in Adults and Kids." last reviewed April 18, 2018. https://www.heart.org/en/healthy-living/fitness /fitness-basics/aha-recs-for-physical-activity-in-adults.

Antonakoudis, G., L. Poulimenos, K. Kifnidis, C. Zouras, and H. Antonakoudis. "Blood Pressure Control and Cardiovascular Risk Reduction," *Hippokratia* 11, no. 3 (July 2007): 114–19.

Aquilina, O. "A Brief History of Cardiac Pacing," *Images in Paediatric Cardiology* 8, no. 2 (April–June 2006): 17–81.

Arráez-Aybar, Luis-Alfonso, Pedro Navia-Álvarez, Talia Fuentes-Redondo, and José-L. Bueno-López. "Thomas Willis, a Pioneer in Translational Research in Anatomy (on the 350th Anniversary of Cerebri Anatome)," *Journal of Anatomy* 226, no. 3 (March 2015): 289–300.

Barclay, Katie. "Dervorgilla of Galloway (abt 1214–abt 1288)." *Women's History Network.* August 15, 2010. https://womenshistorynetwork.org/dervorgilla-of -galloway-abt-1214-abt-1288/.

Bassett, Molly H. *The Fate of Earthly Things: Aztec Gods and God-Bodies.* Austin: University of Texas Press, 1980.

Bendersky, Gordon. "The Olmec Heart Effigy: Earliest Image of the Human Heart," *Perspectives in Biology and Medicine* 40, no. 3 (Spring 1997): 348–61.

Bianucci, Raffaella, Robert D. Loynes, M. Linda Sutherland, Rudy Lallo, Gemma L. Kay, Philippe Froesch, Mark J. Pallen, Philippe Charlier, and Andreas G. Nerlich. "Forensic Analysis Reveals Acute Decompensation of Chronic Heart Failure in a 3500-Year-Old Egyptian Dignitary," *Journal of Forensic Sciences* 61, no. 5 (September 2016): 1378–81.

Bishop, Timothy, and Vincent M. Figueredo. "Hypertensive Therapy: Attacking the Renin-Angiotensin System," *Western Journal of Medicine* 175, no. 2 (August 2001): 119–24.

Burnett, Dean. "Why Elderly Couples Often Die Together: The Science of Broken Hearts." *Guardian,* January 9, 2015. https://www.theguardian.com /lifeandstyle/shortcuts/2015/jan/09/why-elderly-couples-die-together -science-broken-hearts.

Cambiaghi, Marco, and Heidi Hausse. "Leonardo da Vinci and His Study of the Heart," *European Heart Journal* 40, no. 23 (2019): 1823–26.

Camporesi, Piero. *The Incorruptible Flesh: Bodily Mutation and Mortification in Religion and Folklore.* New York: Cambridge University Press, 1988.

Centers for Disease Control and Prevention. "Disparities in Premature Deaths from Heart Disease." February 19, 2004. https://www.cdc.gov/mmwr/preview /mmwrhtml/mm5306a2.htm.

———. "Heart Disease Facts." February 7, 2022. https://www.cdc.gov/heart disease/facts.htm.

———. "Preventing 1 Million Heart Attacks and Strokes." September 6, 2018. https://www.cdc.gov/vitalsigns/million-hearts/.

Chand Tara, Meng Li, Hamidreza Jamalabadi, Gerd Wagner, Anton Lord, Sarah Alizadeh, Lena V. Danyeli, Luisa Herrmann, Martin Walter, and Zumrut D. Sen. "Heart Rate Variability as an Index of Differential Brain Dynamics at Rest and After Acute Stress Induction," *Frontiers in Neuroscience* 14 (July 2020): 645.

Cheang, U. Kei, and Min Jun Kim. "Self-Assembly of Robotic Micro- and Nanoswimmers Using Magnetic Nanoparticles," *Journal of Nanoparticle Research* 17 (2015): 145.

Cingolani, Eugenio, Joshua I. Goldhaber, and Eduardo Marbán. "Next-Generation Pacemakers: From Small Devices to Biological Pacemakers," *Nature Reviews Cardiology* 15, no. 3 (2018): 139–50.

Coding in Tune. "Most Used Words in Lyrics by Genre." April 2018. https://codingintune.com/2018/04/09/statistics-most-used-words-in-lyrics-by-genre/.

Coe, Michael D., and Rex Koontz. *Mexico: From the Olmecs to the Aztecs.* London: Thames & Hudson, 2008.

Cohn, Lawrence H. "Fifty Years of Open-Heart Surgery," *Circulation* 107, no. 17 (2003): 2168–70.

Connelly, Dawn. "A History of Aspirin. *Pharmaceutical Journal.* September 2014. https://pharmaceutical-journal.com/article/infographics/a-history-of-aspirin.

Conradt, Stacey. "Mary Shelley's Favorite Keepsake: Her Dead Husband's Heart." *Mental Floss.* July 8, 2015. https://www.mentalfloss.com/article/65624/mary-shelleys-favorite-keepsake-her-dead-husbands heart.

Critchley Hugo D., and Sarah N. Garfinkel. "Interoception and Emotion," *Currnet Opinion in Psychology* 17 (April 2017): 7–14.

Dalley, Stephanie. *Myths from Mesopotamia: Creation, the Flood, Gilgamesh, and Others.* Oxford: Oxford University Press, 1989.

Diaz Del Castillo, Bernal. *The True History of the Conquest of New Spain.* London: Penguin Classics, 2003.

Díaz-Navarro, Rienzi. "Takotsubo Syndrome: The Broken-Heart Syndrome," *British Journal of Cardiology* 28 (2021): 30–34.

Edriss, Hawa, Brittany N. Rosales, Connie Nugent, Christian Conrad, and Kenneth Nugent. "Islamic Medicine in the Middle Ages," *American Journal of the Medical Sciences* 354, no. 3 (September 2017): 223–29.

Emery, Michael S., and Richard J. Kovacs. "Sudden Cardiac Death in Athletes," *JACC Heart Failure* 6, no. 1 (2018): 30–40.

Enos, W. F., R. H. Holmes, and J. Beyer. "Coronary Disease Among United States Soldiers Killed in Action in Korea: Preliminary Report," *JAMA* 152, no. 12 (1953): 1090–93.

Faulkner, Raymond Oliver. *The Ancient Egyptian Book of the Dead.* London: British Museum Press, 2010.

Fernández-Ruiz, Irene. "Breakthrough in Heart Xenotransplantation," *Nature Reviews Cardiology* 16, no. 2 (February 2019): 69.

Ferrari, Roberto, Cristina Balla, and Alessandro Fucili. "Heart Failure: An Historical Perspective," *European Heart Journal Supplements* 18 (Suppl. G, 2016): G3–G10.

Figueredo, Vincent M. "The Time Has Come for Physicians to Take Notice: The Impact of Psychosocial Stressors on the Heart," *American Journal of Medicine* 122, no. 8 (2009): 704–12.

Fuchs, Thomas. *Mechanization of the Heart: Harvey and Descartes.* Rochester, NY: University of Rochester Press, 2001.

Fye, W. Bruce. "A History of the Origin, Evolution, and Impact of Electrocardiography," *American Journal of Cardiology* 73, no. 13 (1994): 937–49.

——. "Profiles in Cardiology: René Descartes," *Clinical Cardiology* 26, no. 1 (2003): 49–51.

Gabisonia K., G. Prosdocimo, G. D. Aquaro, L. Carlucci, L. Zentilin, I. Secco, H. Ali, L. Braga, N. Gorgodze, F. Bernini, S. Burchielli, C. Collesi, L. Zandonà, G. Sinagra, M. Piacenti, S. Zacchigna, R. Bussani, F. A. Recchia, and M. Giacca. "MicroRNA Therapy Stimulates Uncontrolled Cardiac Repair After Myocardial Infarction in Pigs," *Nature* 569, no. 7756 (2019): 418–22.

Garfinkel, Sarah. "It's an Intriguing World That Is Opening Up," *The Psychologist* 32 (January 2019): 38–41.

Global Observatory on Donation and Transplantation. "Total Heart." April 3, 2022. http://www.transplant-observatory.org/data-charts-and-tables/chart/.

Go, Alan S., Dariush Mozaffarian, Véronique L. Roger, Emelia J. Benjamin, Jarett D. Berry, William B. Borden, Dawn M. Bravata, Shifan Dai, Earl S. Ford, Caroline S. Fox, Sheila Franco, Heather J. Fullerton, Cathleen Gillespie, Susan M. Hailpern, John A. Heit, Virginia J. Howard, Mark D. Huffman, Brett M. Kissela, Steven J. Kittner, Daniel T. Lackland, Judith H. Lichtman, Lynda D. Lisabeth, David Magid, Gregory M. Marcus, Ariane Marelli, David B. Matchar, Darren K. McGuire, Emile R. Mohler, Claudia

S. Moy, Michael E. Mussolino, Graham Nichol, Nina P. Paynter, Pamela J. Schreiner, Paul D. Sorlie, Joel Stein, Tanya N. Turan, Salim S. Virani, Nathan D. Wong, Daniel Woo, and Melanie B. Turner. "Heart Disease and Stroke Statistics—2013 Update: A Report from the American Heart Association," *Circulation* 127, no. 1 (January 2013): e6–e245.

Grossman M., S. E. Raper, K. Kozarsky, E. A. Stein, J. F. Engelhardt, D. Muller, P. J. Lupien, and J. M. Wilson. "Successful Ex Vivo Gene Therapy Directed to Liver in a Patient with Familial Hypercholesterolaemia," *Nature Genetics* 6, no. 4 (1994): 335–41.

Hajar, Rachel. "Al-Razi: Physician for All Seasons," *Heart Views* 6, no. 1 (2005): 39–43.

——. "Coronary Heart Disease: From Mummies to 21st Century," *Heart Views* 18, no. 2 (2017): 68–74.

Harris, C. R. S. *The Heart and Vascular System in Ancient Greek Medicine: From Alcmaeon to Galen.* Oxford: Oxford University Press, 1973.

Haverford College, Intro to Environmental Anthropology Class. "The Gwich'in People: Caribou Protectors." December 2021. https://anthro281 .netlify.app.

Heberden, William. "Some Account of a Disorder of the Breast," *Medical Transactions. The Royal College of London* 2 (1772): 59–67.

Hemakom, Apit, Katarzyna Powezka, Valentin Goverdovsky, Usman Jaffer, and Danilo P. Mandic. "Quantifying Team Cooperation Through Intrinsic Multi-Scale Measures: Respiratory and Cardiac Synchronization in Choir Singers and Surgical Teams," *Royal Society Open Access* 4, no. 12 (November 2017): 170853.

Herrick, James B. "Clinical Features of Sudden Obstruction of the Coronary Arteries," *JAMA* 59 (1912): 2015–20.

Higuchi, Akon, Nien-Ju Ku, Yeh-Chia Tseng, Chih-Hsien Pan, Hsing-Fen Li, S. Suresh Kumar, Qing-Dong Ling, Yung Chang, Abdullah A. Alarfaj, Murugan A. Munusamy, Giovanni Benelli, and Kadarkarai Muruga. "Stem Cell Therapies for Myocardial Infarction in Clinical Trials: Bioengineering and Biomaterial Aspects," *Laboratory Investigation* 97 (2017): 1167–79.

Huang, Shixing, Yang Yang, Qi Yang, Qiang Zhao, and Xiaofeng Ye. "Engineered Circulatory Scaffolds for Building Cardiac Tissue," *Journal of Thoracic Disease* 10 (Suppl. 20, 2018): S2312–28.

Iop, Laura, Eleonora Dal Sasso, Roberta Menabò, Fabio Di Lisa, and Gino Gerosa. "The Rapidly Evolving Concept of Whole Heart Engineering," *Stem Cells International* (2017): 8920940.

James, William. *The Principles of Psychology*. New York: Henry Holt, 1890.

Jayaram, V. "The Meaning and Significance of Heart in Hinduism." 2019. https://www.hinduwebsite.com/hinduism/essays/the-meaning-and -significance-of-heart-in-hinduism.asp.

Kalin, Adi. "Frau Minne hat sich gut gehalten." *NZZ*. November 25, 2009. https:// www.nzz.ch/frau_minne_hat_sich_gut_gehalten-ld.930946?reduced=true.

Keller Andreas, Angela Graefen, Markus Ball, Mark Matzas, Valesca Bois-guerin, Frank Maixner, Petra Leidinger, Christina Backes, Rabab Khairat, Michael Forster, Björn Stade, Andre Franke, Jens Mayer, Jessica Spangler, Stephen McLaughlin, Minita Shah, Clarence Lee, Timothy T. Harkins, Alexander Sartori, Andres Moreno-Estrada, Brenna Henn, Martin Sikora, Ornella Semino, Jacques Chiaroni, Siiri Roostsi, Natalie M. Myres, Vicente M. Cabrera, Peter A. Underhill, Carlos D. Bustamante, Eduard Egarter Vigl, Marco Samadelli, Giovanna Cipollini, Jan Haas, Hugo Katus, Brian D. O'Connor, Marc R. J. Carlson, Benjamin Meder, Nikolaus Blin, Eck-art Meese, Carsten M. Pusch, and Albert Zink. "New Insights Into the Tyrolean Iceman's Origin and Phenotype as Inferred by Whole-Genome Sequencing," *Nature Communications* 3 (February 2012): 698.

King, Edmund. "Arthur Coga's Blood Transfusion (1667)." *Public Domain Review*. April 15, 2014. https://publicdomainreview.org/collection/arthur -coga-s-blood-transfusion-1667.

King, Helen. *Greek and Roman Medicine*. London: Bristol Classical Press, 2001.

Längin, Martha, Tanja Mayr, Bruno Reichart, Sebastian Michel, Stefan Buch-holz, Sonja Guethoff, Alexey Dashkevich, Andrea Baehr, Stephanie Egerer, Andreas Bauer, Maks Mihalj, Alessandro Panelli, Lara Issl, Jiawei Ying, Ann Kathrin Fresch, Ines Buttgereit, Maren Mokelke, Julia Radan, Fabian Werner, Isabelle Lutzmann, Stig Steen, Trygve Sjöberg, Audrius Paskev-icius, Liao Qiuming, Riccardo Sfriso, Robert Rieben, Maik Dahlhoff, Barbara Kessler, Elisabeth Kemter, Mayulko Kurome, Valeri Zakhartch-enko, Katharina Klett, Rabea Kingel, Christian Kupatt, Almuth Falkenau, Simone Reu, Reinhrad Ellgass, Rudolf Herzog, Uli Binder, Günter Wich, Arne Skerra, David Ayares, Alexander Kind, Uwe Schönmann, Franz-Josef Kaup, Christain Hagl, Eckhard Wolf, Nikolai Klymuk, Paolo Brenner, and Jan-Michael Abicht. "Consistent Success in Life-Supporting Porcine Car-diac Xenotransplantation," *Nature* 564, no. 7736 (2018): 430–33.

Lee, Moo-Sik, Andreas J. Flammer, Lilach O. Lerman, and Amir Lerman. "Personalized Medicine in Cardiovascular Diseases," *Korean Circulation Journal* 42, no. 9 (September 2012): 583–91.

REFERENCES

Lillehei, C. W. "The Society Lecture. European Society for Cardiovascular Surgery Meeting, Montpellier, France, September 1992. The Birth of Open-Heart Surgery: Then the Golden Years," *Cardiovascular Surgery* 2, no. 3 (1994): 308–17.

Loukas Marios, Pamela Clarke, R. Shane Tubbs, and Theodoros Kapos. "Raymond de Vieussens," *Anatomical Science International* 82, no. 4 (2007): 233–36.

Ma, Xiaoya, Peiyun Cong, Xianguang Hou, Gregory D. Edgecombe, and Nicholas J. Strausfeld. "An Exceptionally Preserved Arthropod Cardiovascular System from the Early Cambrian," *Nature Communications* 5 (2014): 3560.

Maas A. H. E. M., and Y. E. A. Appelman. "Gender Differences in Coronary Heart Disease," *Netherlands Heart Journal* 18, no. 12 (December 2010): 598–602.

Macrobius, Ambrosius Aurelius Theodosius. *Seven Books of the Saturnalia.* accessed April 2022. https://www.loc.gov/item/2021667911/.

Mafart, Bertrand. "Post-Mortem Ablation of the Heart: A Medieval Funerary Practice. A Case Observed at the Cemetery of Ganagobie Priory in the French Department of Alpes De Haute Provence," *International Journal of Osteoarchaeology* 14, no. 1 (2004): 67–73.

Maffie, James. "Aztec Philosophy." *Internet Encyclopedia of Philosophy.* April 3, 2022. https://iep.utm.edu/aztec-philosophy/.

Maher, Brendan. "Tissue Engineering: How to Build a Heart," *Nature* 499 (2013): 20–22.

McCraty Rollin, Mike Atkinson, Dana Tomasino, and Raymond Trevor Bradley. "The Coherent Heart: Heart–Brain Interactions, Psychophysiological Coherence, and the Emergence of System-Wide Order," *Integral Review* 5, no. 2 (December 2009): 10–115.

McNamara, J. J., M. A. Molot, J. F. Stremple, and R. T. Cutting. "Coronary Artery Disease in Combat Casualties in Vietnam," *JAMA* 216, no. 7 (1971): 1185–87.

Miller, Michael. "Emotional Rescue: The Heart-Brain Connection," *Cerebrum* (May 2019): cer-05-19.

Miller, T. Christian. "A History of the Purple Heart." *NPR.* September 2010. https://www.npr.org/templates/story/story.php?storyId=129711544.

Miner, Jonathan, and Adam Hoffhines. "The Discovery of Aspirin's Antithrombotic Effects," *Texas Heart Journal* 34, no. 2 (2007): 179–86.

Modelli, Manoel E. S., Aurea S. Cherulli, Lenora Gandolfi, and Riccardo Pratesi. "Atherosclerosis in Young Brazilians Suffering Violent Deaths: A Pathological Study," *BMC Research Notes* 4 (2011): 531.

Montinari, Maria Rosa, and Sergio Minelli. "The First 200 Years of Cardiac Auscultation and Future Perspectives," *Journal of Multidisciplinary Healthcare* 12 (2019): 183–89.

Musunuru, Kiran, Alexandra C. Chadwick, Taiji Mizoguchi, Sara P. Garcia, Jamie E. DeNizio, Caroline W. Reiss, Kui Wang, Sowmya Iyer, Chaitali Dutta, Victoria Clendaniel, Michael Amaonye, Aaron Beach, Kathleen Berth, Souvik Biswas, Maurine C. Braun, Huei-Mei Chen, Thomas V. Colace, John D. Ganey, Soumyashree A. Gangopadhyay, Ryan Garrity, Lisa N. Kasiewicz, Jennifer Lavoie, James A. Madsen, Yuri Matsumoto, Anne Marie Mazzola, Yusuf S. Nasrullah, Joseph Nneji, Huilan Ren, Athul Sanjeev, Madeleine Shay, Mary R. Stahley, Steven H. Y. Fan, Ying K. Tam, Nicole M. Gaudelli, Giuseppe Ciaramella, Leslie E. Stolz, Padma Malyala, Christopher J. Cheng, Kallanthottathil G. Rajeev, Ellen Rohde, Andrew M. Bellinger, and Sekar Kathiresan. "In Vivo CRISPR Base Editing of PCSK9 Durably Lowers Cholesterol in Primates," *Nature* 593, no. 7859 (2021): 429–34.

Nature Editors. "Samuel Siegfried Karl von Basch (1837–1905)," *Nature* 140 (1937): 393–94.

Neferet, Amentet. *Ancient Egyptian Dictionary.* accessed December 2021. https://seshkemet.weebly.com/dictionary.html.

Nunn, John F. *Ancient Egyptian Medicine.* London: British Museum Press, 1996.

Obrastzow, W. P., and N. D. Staschesko. "Zur Kenntnissder Thrombose der Coronararterien des Herzens," *Zeitschrift für klinische Medizin* 71 (1910): 12.

O'Rourke Boyle, Marjorie. "Aquinas's Natural Heart," *Early Science and Medicine* 18, no. 3 (2013): 266–90.

——. *Cultural Anatomies of the Heart in Aristotle, Augustine, Aquinas, Calvin, and Harvey.* London: Palgrave Macmillan, 2018.

Patwardhan, Kishor. "The History of the Discovery of Blood Circulation: Unrecognized Contributions of Ayurveda Masters," *Advances in Physiology Education* 36, no. 2 (2012): 77–82.

Paul. "Door 23: The Heart of a King." *Geological Society of London* (blog). December 23, 2014. https://blog.geolsoc.org.uk/2014/12/23/the-heart-of-a-king/.

Perciaccante A., M. A. Riva, A. Coralli, P. Charlier, and R. Bianucci. "The Death of Balzac (1799–1850) and the Treatment of Heart Failure During the Nineteenth Century," *Journal of Cardiac Failure* 22, no. 11 (2016): 930–33.

Pettinari, Matteo, Emiliano Navarra, Philippe Noirhomme, and Herbert Gutermann. "The State of Robotic Cardiac Surgery in Europe," *Annals of Cardiothoracic Surgery* 6, no. 1 (2017): 1–8.

Prieto, Gabriel, John W. Verano, Nicolas Goepfert, Douglas Kennett, Jeffrey Quilter, Steven LeBlanc, Lars Fehren-Schmitz, Jannine Forst, Mellisa Lund, Brittany Dement, Elise Dufour, Olivier Tombret, Melina Calmon, Davette Gadison, and Khrystyne Tschinkel. "A Mass Sacrifice of Children and Camelids at the Huanchaquito-Las Llamas Site, Moche Valley, Peru," *PLoS One* 14, no. 3 (2019): e0211691.

Raal, Frederick J., David Kallend, Kausik K. Ray, Traci Turner, Wolfgang Koenig, R. Scott Wright, Peter L. J. Wijngaard, Danielle Curcio, Mark J. Jaros, Lawrence A. Leiter, John J. P. Kastelein, and ORION-9 Investigators. "Inclisiran for the Treatment of Heterozygous Familial Hypercholesterolemia," *New England Journal of Medicine* 382, no. 16 (2020): 1520–30.

Ranhel, André Silva, and Evandro Tinoco Mesquita. "The Middle Ages Contributions to Cardiovascular Medicine," *Brazilian Journal of Cardiovascular Surgery* 31, no. 2 (April 2016): 163–70.

Reveron, Rafael Romero. "Herophilus and Erasistratus, Pioneers of Human Anatomical Dissection," *Vesalius* 20, no. 1 (2014): 55–58.

Roguin, Ariel. "Rene Theophile Hyacinthe Laënnec (1781–1826): The Man Behind the Stethoscope," *Clinical Medicine and Research* 4, no. 3 (2006): 230–35.

Rosengren Annika, Steven Hawken, Stephanie Ounpuu, Karen Sliwa, Mohammad Zubaid, Wael A. Almahmeed, Kathleen Ngu Blackett, Chitr Sitthi-amorn, Hiroshi Sato, Salim Yusuf, and INTERHEART investigators. "Association of Psychosocial Risk Factors with Risk of Acute Myocardial Infarction in 11119 Cases and 13648 Controls from 52 Countries (the INTERHEART study): Case-Control Study," *Lancet* 364, no. 9438 (2004): 953–62.

Roser, Max, Esteban Ortiz-Ospina, and Hannah Ritchie. "Life Expectancy." *Our World in Data.* October 2019. https://ourworldindata.org/life-expectancy.

Sakatani, Kaoru. "Concept of Mind and Brain in Traditional Chinese Medicine," *Data Science Journal* 6 (Suppl., 2007): S220–24.

Sanders, N. K. *The Epic of Gilgamesh*. London: Penguin Books, 1972.

Shaffer Fred, Rollin McCraty, and Christopher L. Zerr. "A Healthy Heart Is Not a Metronome: An Integrative Review of the Heart's Anatomy and Heart Rate Variability," *Frontiers in Psychology* 5 (2014): 1040.

Shah, Mahek, Pradhum Ram, Kevin Bryan U. Lo, Natee Sirinvaravong, Brijesh Patel, Byomesh Tripathi, Shantanu Patil, and Vincent M. Figueredo. "Etiologies, Predictors, and Economic Impact of Readmission Within 1

Month Among Patients with Takotsubo Cardiomyopathy," *Clinical Cardiology* 41, no. 7 (July 2018): 916–23.

Silverman, Mark E. "Andreas Vesalius and de Humani Corporis Fabrica," *Clinical Cardiology* 14 (1991): 276–79.

Slights, William W. E. "The Narrative Heart of the Renaissance," *Renaissance and Reformation* 26, no. 1 (2002): 5–23.

Song, L. Z., G. E. Schwartz, and L. G. Russek. "Heart-Focused Attention and Heart-Brain Synchronization: Energetic and Physiological Mechanisms," *Alternative Therapies in Health and Medicine* 4, no. 5 (September 1998): 44–52, 54–60, 62.

Specktor, Brandon. "Evolution Turned This Fish Into a 'Penis with a Heart.' Here's How." *Live Science.* August 3, 2020. https://www.livescience.com /anglerfish-fusion-sex-immune-system.html.

Sturlason, Snorre. *Heimskringla—The Norse King Sagas.* Redditch, UK: Read Books, 2008.

Swann, Jeremy B., Stephen J. Holland, Malte Petersen, Theodore W. Pietsch, and Thomas Boehm. "The Immunogenetics of Sexual Parasitism," *Science* 369, no. 6511 (2020): 1608–15.

Thayer Julian F., and Richard D. Lane. "Claude Bernard and the Heart-Brain Connection: Further Elaboration of a Model of Neurovisceral Integration," *Neuroscience & Biobehavioral Reviews* 33, no. 2 (2009): 81–88.

Thompson Randall C., Adel H. Allam, Guido P. Lombardi, L. Samuel Wann, M. Linda Sutherland, James D. Sutherland, Muhammad Al-Tohamy Soliman, Bruno Frohlich, David T. Mininberg, Janet M. Monge, Clide M. Vallodolid, Samantha L. Cox, Gomaa Abd el-Maksoud, Ibrahim Badr, Michael I. Miyamoto, Abd el-Halim Nur el-din, Jagat Narula, Caleb E. Finch, and Gregory S. Thomas. "Atherosclerosis Across 4000 Years of Human History: The Horus Study of Four Ancient Populations," *Lancet* 381, no. 9873 (2013): 1211–22.

Tofield, A. "Hikaru Sato and Takotsubo Cardiomyopathy," *European Heart Journal* 37, no. 37 (October 2016): 2812.

Toro, Ross. *Leading Causes of Death in the US: 1900–Present* (Infographic). July 1, 2012. https://www.livescience.com/21213-leading-causes-of-death-in -the-u-s-since-1900-infographic.html.

Veith, Ilza. *Huang Ti Nei Ching Su Wen: The Yellow Emporer's Classic of Internal Medicine.* Baltimore, MD: Williams & Wilkins, 1949.

U.S. Department of Health and Human Services. "Heart Disease and African Americans." January 31, 2022. https://minorityhealth.hhs.gov/omh/browse aspx?lvl=4&lvlid=19

REFERENCES

Vickhoff, Björn, Helge Malmgren, Rickard Aström, Gunnar Nyberg, Seth-Reino Ekström, Mathias Engwall, Johan Snygg, Michael Nilsson, and Rebecka Jörnsten. "Music Structure Determines Heart Rate Variability of Singers," *Fronties in Psychology* 4 (July 2013): 334.

Vinken, Pierre. "How the Heart Was Held in Medieval Art," *Lancet* 358, no. 9299 (2001): 2155–57.

Vogenberg, F. Randy, Carol Isaacson Barash, and Michael Pursel. "Personalized Medicine: Part 1: Evolution and Development Into Theranostics," *Pharmacy and Therapeutics* 35, no. 10 (2010): 565–67.

Wasfy, Meagan M., Adolph M. Hutter, and Rory B. Weiner. "Sudden Cardiac Death in Athletes," *Methodist Debakey Cardiovascular Journal* 12, no. 2 (2016): 76–80.

Webb, Heather. *The Medieval Heart*. New Haven, CT: Yale University Press, 2010.

West, John B. "Marcello Malpighi and the Discovery of the Pulmonary Capillaries and Alveoli," *American Journal of Physiology, Lung Cellular and Molecular Physiology* 304, no. 6 (2013): L383–90.

World Health Organization. "Cardiovascular Diseases." June 2021. https://www.who.int/news-room/fact-sheets/detail/cardiovascular-diseases-(cvds).

——. "Hypertension." August 25, 2021. https://www.who.int/news-room/fact-sheets/detail/hypertension.

——. "The Top 10 Causes of Death." December 2020. https://www.who.int/news-room/fact-sheets/detail/the-top-10-causes-of-death.

Yu, Jiangfan, Ben Wang, Xingzhou Du, Qianqian Wang, and Li Zhang. "Ultra-Extensible Ribbon-Like Magnetic Microswarm," *Nature Communications* 9, no. 1 (2018): 3260.

Zysk, Kenneth G. *Religious Medicine: History and Evolution of Indian Medicine*. London: Transaction, 1993.

FURTHER READING

BOOKS

Acierno, Louis J. "Physical Examination." In *The History of Cardiology*, 447–492. London: Parthenon, 1994.

Amidon, Stephen, and Thomas Amidon. *The Sublime Engine: A Biography of the Human Heart*. New York: Rodale, 2011.

Boyadjian, N. *The Heart: Its History, Its Symbolism, Its Iconography and Its Diseases*. Antwerp: Esco, 1985.

Celsus, A. Cornelius. *On Medicine, Volume I: Books 1–4*. trans. W. G. Spencer. Cambridge, MA: Harvard University Press, 1935.

Dunn, Rob. *The Man Who Touched His Own Heart: True Tales of Science, Surgery, and Mystery*. New York: Little, Brown, 2015.

Fishman, Alfred P., and Dickinson W. Richards. *Circulation of the Blood: Men and Ideas*. New York: Springer, 1982.

Forrester, James. *The Heart Healers: The Misfits, Mavericks, and Rebels Who Created the Greatest Medical Breakthrough of Our Lives*. New York: St. Martin's Press, 2015.

Harvey, William. *An Anatomical Disquisition on the Motion of the Heart and Blood in Animals*. trans. Robert Willis. London: Dent, 1907.

Homer. *The Iliad*. London: Penguin Classics, 1998.

Høystad, Ole M. *A History of the Heart*. London: Reaktion, 2007.

Jauhar, Sandeep. *Heart: A History*. New York: Farrar, Straus, and Giroux, 2018.

Larrington, Carolyne. *The Poetic Edda*. Oxford: Oxford World's Classics, 1936.

Lloyd, G. E. R. *Hippocratic Writings*. London: Penguin, 1978.

McCrae, Donald. *Every Second Counts: The Race to Transplant the First Human Heart*. New York: Putnam, 2006.

Monagan, David. *Journey Into the Heart: A Tale of Pioneering Doctors and Their Race to Transform Cardiovascular Medicine*. New York: Gotham, 2007.

Slights, William W. E. *The Heart in the Age of Shakespeare*. New York: Cambridge University Press, 2008.

Smith, J. V. C., ed. *The Boston Medical and Surgical Journal. Volume XXVII*. Boston: D. Clapp Jr., 1843.

Smith, Michael E. *The Aztecs*. Malden, MA: Blackwell, 2003.

Warraich, Haider. *State of the Heart: Exploring the History, Science, and Future of Cardiac Disease*. New York: St. Martin's Press, 2019.

ARTICLES

Aird, W. C. "Discovery of the Cardiovascular System: From Galen to William Harvey," *Journal of Thrombosis Haemostasis* 9 (2011): 118–29.

Al Ghatrif, Majd, and Joseph Lindsay. "A Brief Review: History to Understand Fundamentals of Electrocardiography," *Journal of Community Hospital Internal Medicine Perspectives* 2, no. 1 (2012): 14383.

Benjamin, Emelia J., et al. "Heart Disease and Stroke Statistics—2018 Update: A Report from the American Heart Association," *Circulation* 137 (2018): e67–e492.

Besser, Michael. "Galen and the Origins of Experimental Neurosurgery," *Austin Journal of Surgery* 1, no. 2 (2014): 1009.

Boon, Brigitte. "Leonardo da Vinci on Atherosclerosis and the Function of the Sinuses of Valsalva," *Netherlands Heart Journal* 17, no. 12 (2009): 496–99.

Braunwald, Eugene. "Cardiology: The Past, the Present, and the Future," *JAMA* 42, no. 12 (2003): 2031.

Cooley, Denton A. "Some Thoughts About the Historical Events That Led to the First Clinical Implantation of a Total Artificial Heart," *Texas Heart Institute Journal* 40 (2013): 117–19.

Eknoyan, Garabed. "Emergence of the Concept of Cardiovascular Disease," *Advances in Chronic Kidney Disease* 11, no. 3 (2004): 304–9

Forssmann-Falck, Renate. "Werner Forssmann: A Pioneer of Cardiology," *American Journal of Cardiology* 79 (1997): 651–60.

French, R. K. "The Thorax in History. 1. From Ancient Times to Aristotle," *Thorax* 33 (1978): 10–18.

French, R. K. "The Thorax in History. 2. Hellenistic Experiment and Human Dissection," *Thorax* 33 (1978): 153–66.

Fye, W. Bruce. "Lauder Brunton and Amyl Nitrite: A Victorian Vasodilator," *Circulation* 74 (1986): 222–29.

Ghosh, Sanjib K. "Human Cadaveric Dissection: A Historical Account from Ancient Greece to the Modern Era," *Anatomy & Cell Biology* 48, no. 3 (2015): 153–69.

Gilbert, N. C. "History of the Treatment of Coronary Heart Disease," *JAMA* 148, no. 16 (1952): 1372–76.

Hajar, Rachel. "The Pulse in Ancient Medicine—Part 1," *Heart Views* 19 (2018): 36–43.

Heron, Melonie. "Deaths: Leading Causes for 2017," *National Vital Statistics Reports* 68, no. 6 (2019): 1–76.

Herrick, James. "An Intimate Account of My Early Experience with Coronary Thrombosis," *American Heart Journal* 27 (1944): 1–18.

Lonie, I. M. "The Paradoxical Text 'on the Heart,' Part 1," *Medical History* 17 (2012): 1–15.

Madjid, Mohammad, Payam Safavi-Naeini, and Robert Loder. "High Prevalence of Cholesterol-Rich Atherosclerotic Lesions in Ancient Mummies: A Near-Infrared Spectroscopy Study," *American Heart Journal* 216 (2019): 113–16.

Miller, Leslie W., and Joseph G. Rogers. "Evolution of Left Ventricular Assist Device Therapy for Advanced Heart Failure," *JAMA Cardiology* 3, no. 7 (2018): 650–58.

Muller, James E. "Diagnosis of Myocardial Infarction: Historical Notes from the Soviet Union and the United States," *American Journal of Cardiology* 40 (1977): 269–71.

Murphy, Sherry L., Jiaquan Xu, and Kenneth D. Kochanek. "Deaths: Final Data for 2010," *National Vital Statistics Reports* 61, no. 4 (2013).

Meyers, Jonathan. "Exercise and Cardiovascular Health," *Circulation* 107, no. 1 (2003): e2–5.

Park, Katherine. "The Life of the Corpse: Division and Dissection in Late Medieval Europe," *Journal of the History of Medicine and Allied Sciences* 50, no. 1 (1995): 111–32.

Pasipoularides, Ares. "Galen, Father of Systematic Medicine. An Essay on the Evolution of Modern Medicine and Cardiology," *International Journal of Cardiology* 172 (2014): 47–58.

Reynolds, Edward H., and James V. Kinnier Wilson. "Neurology and Psychiatry in Babylon," *Brain* 137, no. 9 (2014): 2611–19.

Saba, Magdi M., Hector O. Ventura, Mohamed Saleh, and Mandeep R. Mehra. "Ancient Egyptian Medicine and the Concept of Heart Failure," *Journal of Cardiac Failure* 12 (2006): 416–21.

Schultz, Stanley G. "William Harvey and the Circulation of the Blood: The Birth of a Scientific Revolution and Modern Physiology," *Physiology* 17, no. 5 (2002): 175–80.

Shoja, Mohammadali M., Paul S. Agutter, Marios Loukas, Brion Benninger, Ghaffar Shokouhi, Husain Namdar, Kamyar Ghabili, Majid Khalili, and R. Shane Tubbs. "Leonardo da Vinci's Studies of the Heart," *International Journal of Cardiology* 167, no. 4 (2013): 1126–33.

Sterpetti, Antonio V. "Cardiovascular Research by Leonardo da Vinci," *Circulation Research* 2124 (2019):189–91.

Thiene, Gaetano, and Jeffrey E. Saffitz. "Response by Thiene and Saffitz to Letter Regarding Article, 'Autopsy as a Source of Discovery in Cardiovascular Medicine: Then and Now,'" *Circulation* 139, no. 4 (2019): 568–69.

Thomas, Gregory S., et al. "Why Did Ancient People Have Atherosclerosis? From Autopsies to Computed Tomography to Potential Causes," *Global Heart* 9, no. 2 (2014): 229–37.

Uddin, Lucina Q., Jason S. Nomi, Benjamin Hébert-Seropian, Jimmy Ghaziri, and Olivier Boucher. "Structure and Function of the Human Insula," *Journal of Clinical Neurophysiology* 34, no. 4 (2017): 300–306.

Vinaya, P. N., and J. S. R. A. Prasad. "The Concept of Blood Circulation in Ancient India W.S.R. to the Heart as a Pumping Organ," *International Ayurvedic Medical Journal* 2, no. 15 (2015): 244–49.

Willerson James T., and Rebecca Teaff. "Egyptian Contributions to Cardiovascular Medicine," *Texas Heart Institute Journal* 23 (1996): 191–200.

WEB SOURCES

Dharmananda, Subhuti. "The Significance of Traditional Pulse Diagnosis in the Modern Practice of Chinese Medicine." *Institute for Traditional Medicine.* August 2000. http://www.itmonline.org/arts/pulse.htm.

Elliott, Martin, and Valerie Shrimplin. "Affairs of the Heart: An Exploration of the Symbolism of the Heart in Art." *Gresham College*. February 14, 2017. https://www.gresham.ac.uk/lectures-and-events/affairs-of-the-heart-an -exploration-of-the-symbolism-of-the-heart-in-art.

Institute for Traditional Medicine. "The Heart: Views from the Past." accessed April 3, 2022. http://www.itmonline.org/5organs/heart.htm.

Love, Shayla. "Can You Feel Your Heartbeat? The Answer Says a Lot About You." *Vice*. February 3, 2020. https://www.vice.com/en/article/akw3xb/connection -between-heartbeat-anxiety.

Rosch, Paul J. "Why the Heart Is Much More Than a Pump." *HeartMath Library Archives*. 2015. https://www.heartmath.org/research/research-library/relevant/ heart-much-pump/.

Wikipedia, The Free Encyclopedia. "Chandogya Upanishad." last edited May 16, 2020. https://en.wikipedia.org/w/index.php?title=Chandogya _Upanishad&oldid=956991823.

INDEX

Page numbers in *italics* indicate illustrations.